The Origin of Species
Revisited

The Origin of Species Revisited

A Victorian Who Anticipated Modern Developments in Darwin's Theory

DONALD R. FORSDYKE

McGill-Queen's University Press
Kingston & Montreal • London • Ithaca

Legal deposit fourth quarter 2001
Bibliothèque nationale du Québec

Printed in Canada on acid-free paper

This book has been published with the help of a grant from the Humanities and Social Sciences Federation of Canada, using funds provided by the Social Sciences and Humanities Research Council of Canada, and with the help of funds from Queen's University.

McGill-Queen's University Press acknowledges the financial support of the Government of Canada through the Book Publishing Industry Development Program (BPIDP) for its activities. It also acknowledges the support of the Canada Council for the Arts for its publishing program.

National Library of Canada Cataloguing in Publication Data

Forsdyke, Donald
    The origin of species revisited: a Victorian who anticipated
    modern developments in Darwin's theory
    Includes bibliographical references and index
    ISBN 0-7735-2259-X
    1. Evolution (Biology) - Great Britain - History - 19th
    century. 2. Romanes, George John, 1848-1894. I. Title.
    QH361.F67 2001   576.8'2'094109034   C2001-900151-7

To Patricia, Ruth, Sara, Polly, and Charlotte

Contents

Acknowledgments

I am indebted to William Provine whose works greatly eased the sifting of the "Darwiniana" literature. Provine (1986 ch.7, 207) admits that "evolutionary biology in the period 1859-1925 is extraordinarily complex." It is my hope that this book has made it less so. Elizabeth Barnes made available her unpublished Cambridge B.A. thesis on Romanes. John Ringereide kindly donated Mabel Ringereide's unpublished papers on the Romanes family, which are in Queen's University Archives. Anne Barrett of the Archives at the Imperial College of Science, Technology and Medicine in London made available correspondence from the Huxley Archives. P. Divakar provided information on the content of Haldane's personal library in Hyderabad.

Permission to reproduce the Haldane drawing was given by the Genetic Society of America. Giles Romanes made available photographs of the portraits of his grandfather and grandmother. Morten Kielland-Brandt provided the photograph of Öjvind Winge, with the permission of the Carlsberg Foundation. The photograph of Darwin as a young man is from an original drawing held by the Fitzwilliam Museum, Cambridge. Figure 9.1 is adapted from Crick's 1971 paper, and Figure 9.2 is adapted from Gierer's 1966 paper, with the permission of the publishers of *Nature*. Figure 10.2 is from Bronson and Andersons' 1994 paper with the permission of Springer-Verlag.

Academic Press gave permission to include parts of papers published in the *Journal of Theoretical Biology*. The editor of *Queen's Quarterly* gave permission for material in chapter 20. Gordon and Breach (Harwood Academic) gave permission to include material in chapter 21 from my book *Tomorrow's Cures Today?*. My bioinformatic studies referred to in Parts 2 and 3 were partly supported by the Medical Research Council of Canada and were enthusiastically assisted by Isabelle Barrette, Sheldon Bell, Labonny Biswas. Yiu Cheung Chow, Anthony Cristillo, Kha Dang, Previn Dutt, Scott Heximer. Gregory Hill, Janet Ho, Perry Lao, James Mortimer, Robert Rasile, and Theresa St. Denis. Much valuable advice and support came from Jim Gerlach, Laverne Russell, and my wife Patricia. Editorial assistance was provided by Geoffrey Smith and Virginia Parker (chapter 20), Jacalyn Duffin (chapter 21), and Charlotte and Ruth Forsdyke. Roger Martin, Joan McGilvray, Rachel Mansfield, Lynda Powell, and other staff of McGill-Queen's Press greatly facilitated the production process.

Queen's University hosts my web pages (http://post.queensu.ca/~forsdyke/homepage.htm), which display full-texts of key scientific papers from the nineteenth century onwards and much supplementary material. Here the reader will find regular updates on new work which appeared after the book went to press. To further supplement the book, I have written short biographies of W. Bateson, E. Chargaff, J.B.S. Haldane, G.J. Mendel, H.J. Muller and G.J. Romanes. Macmillan Reference Ltd. has kindly placed these in both the on-line and paper editions of the *Encyclopedia of Life Sciences* (Forsdyke 2001b-g).

The Origin of Species
Revisited

Prologue

"So far as I can venture to offer an opinion on the matter, the purpose of our being in existence, the highest object that human beings can set before themselves, is not the pursuit of any such chimera as the annihilation of the unknown; but it is simply the unwearied endeavour to remove its boundaries a little further from our sphere of action."

T.H. Huxley (1863 ch.11, 449)

For a biologist, the "highest object" for "unwearied endeavour" is the question of the origin of species. It is a bold endeavour, but one which must, from time to time, be revisited. Without an understanding of the origin of species we cannot understand biological evolution – the light which illuminates genes, the biomedical sciences and much of our existence.

The modern story began in 1859 with the Darwin–Wallace theory of the origin of species by natural selection. From the start there were obvious inconsistencies. Then, at the turn of the nineteenth century came two major developments: first, the publication (and rapid consignment to oblivion) of the Romanes–Gulick theory of the origin of species by *physiological selection*; second, the discovery of the work of Mendel. The attempt by Fisher, Haldane, Wright, and Muller to reconcile Darwin's ideas with Mendelian genetics was described in *The Origins of Theoretical Population Genetics* by historian William Provine who concluded (1971 ch.5, 178): "Thus, with the gap between theoretical models and available observational data so large, population genetics began and continues with a theoretical structure containing obvious internal inconsistencies."

Richard Dawkin's *The Selfish Gene* (1976), which magnificently synthesized the work of William Hamilton (1964) and George Williams (1966), added a fresh perspective to Darwinian ideas. However, the new perspective was largely biological and many inconsistencies remained. Indeed, Williams (1975 preface, v) was convinced "that there is a kind of crisis at hand in evolutionary biology." In 1980 Stephen Gould asked: "Is a new and general theory of evolution emerging?" Then in the 1990s came the genome project and a deluge of DNA sequence data from bio-chemical laboratories. The application of powerful computer programs to the bioinformatic analysis of this data has given the story fresh impetus, as is related here.

The book is in four parts. The first part considers the difficulties the Victorians had with Darwin's theory, and my search for an unrecognized Victorian who might have solved them. I draw particular attention to Huxley's difficulty, the "problem of hybridism," and to the "non-adaptationist" ideas first of George Romanes, and then of William Bateson. In a formidable display of inductive reasoning, paralleled only by that of Darwin himself, Romanes (1897) went as far as anyone in the nineteenth century could have been expected to go in reconciling the inconsistencies. Unfortunately the reconciliation could only be achieved by postulating a "physiological peculiarity" of the reproductive system (ch.3, 52). To some this seemingly vague explanation may have appeared not far short of creationism, and it may not have helped Romanes' credibility that his chief ally was the Reverend John Gulick.

In the second part I argue that the chemical basis of the origin of species by means of "physiological selection" is something biochemists have known about for many decades – the species-dependent component of the base composition of DNA. This is the "holy grail" sought by Romanes and Bateson. A failure of this component to match (complement) between two conjugating individuals is manifest in their progeny as hybrid sterility, an internal barrier to the crossing of species. It was the great simplicity of this model which led me to suspect that one of Darwin's more perspicacious associates might have been much nearer to the solution than is currently acknowledged.

In the third part, I consider some outstanding problems in biology and medicine from this new perspective. The problem of the origin of species is seen as part of the general problem of self/not-self discrimination, which is concerned not only with how species *diverge*, but also with how species *converge* in symbiotic or pathogenic relationships. Divergence and convergence are but two sides of the same coin. I show how understanding the genomes of viruses is helping us understand the origin of species. Conversely, understanding the origin of species is helping us understand the genomes of viruses.

The Victorians "discovered" evolution and, to avoid the pitfalls of paraphrase and mistranslation (Sandler 2000), wherever possible I have let them tell the story. However, like biological species, languages have evolved and diverged and converged. The words "genetic" and "genetically" were used by Huxley at least as early as 1864 (ch.3, 93), but cannot have meant in the Victorian era what they mean to us today. The word "virus," meaning "poison," was used decades before the discovery of the micro-organisms we now call "viruses" (Romano 1993, 115; 1997). Thus, to abundant quotations throughout the text, I have added my own annotations with a view to helping readers, particularly those for whom English is not a first language, understand what Darwin and his contemporaries really meant (or rather, what I think they meant) by what they said. Words in square brackets, which are included between quotation marks, are my own, and should not be attributed to the person quoted. Unless otherwise stated, the use of italics to emphasize a word should be attributed to me. Much of the confusion in the evolution literature is semantic, so I have tried to use plain language and a limited number of clearly defined technical terms. Some of these definitions may be controversial, but it is to be recalled that progress in science (e.g., physics at the turn of the nineteenth century) sometimes requires that conventional definitions be challenged.

In attempting to build on past foundations this book is somewhat like a detective story. Who thought what, and when? In the phrase made popular by the Watergate investigation in the U.S.A.: "What did he know and when did he know it?" To help readers who might like to try to put the pieces together for themselves, in the text I have cited the work of others by name and date, rather than by number as is more conventional in books of this nature. The name-and-date approach is less error-prone. There will undoubtedly be errors in this book, but probably not in the citations. Because page numbering tends to change more than chapter numbering in successive editions of a book, where possible quotations are located both by chapter and page number (e.g., "Darwin 1839 ch.13, 422"). To convey better the historical succession of events, dates in citations usually refer to the date of origin of a work. Thus "de Vries 1889, 215" gives the date of the original German edition by de Vries, but refers to page 215 of Gager's 1910 English translation. The reference list makes this unambiguous.

In places I am drawn to ask questions of a historical nature, which I am not fully qualified to answer. What did he believe, and why did he believe it? Why did he place most weight on this at one time, but place more weight on that, at another time? I have read, but not exhaustively analyzed, the main writings of the major protagonists. Unfortunately many stones remain unturned, and I crave the indulgence of Darwin scholars. My excuse is that as a laboratory scientist I see things from a different perspective, and may have something new to offer. My efforts may point to a neglected path, which the professional historians may then choose to follow. Hopefully, the blend of history and science will lend life to a subject which can all-too-easily be lost in technicalities. Indeed, one cannot really understand a subject without understanding its history, and one cannot really understand its history without understanding the subject. This book, an unabashed "Whig history," attempts to resolve this paradox.

The Whigs, with much complacency, tended to judge historical events as to whether they were in accord with Whig political doctrines. One must, of course, always be cautious, but when the facts have gone far beyond doctrine (e.g., the fact that DNA is the main form in which our genetic characteristics are transmitted), then hindsight judgements can be illuminating. But there is more to it than this. Darwin's removal of creationist blinkers enabled people to look afresh at major evolutionary questions. Some, like Darwin, were very bright, independently wealthy, and had the enthusiasm, resources, and time to think deeply on the subject in a way few can today. The literature was relatively finite, there were no grant applications to write, there were less distractions, and the "major players" were few, yet accessible.

To the extent that the story turns out to have a hero, it is Romanes, with Bateson a close second. Romanes is now best-known for the Oxford lecture series which he endowed, and for his contributions to psychology. In my view Romanes' contribution to evolutionary biology is no less significant than that of Mendel. The latter's work (1865) had to wait thirty-five years before it was appreciated. Romanes' major work (1897) has waited a century. In the fourth part of the book I suggest some reasons for this. As the book neared completion, I was delighted to find *The Life and Letters of George John Romanes* (Romanes 1896) in the Queen's University library, and was much surprised to learn that Romanes was born here in Kingston, Canada, the son of a Queen's University classics professor (Ringereide 1979).

To whom is this book directed? No one in particular. The book was written to clarify and amplify papers on biological evolution which I published in the 1990s, and which I felt needed to be brought together as one text to support my web-pages on the topic. The pages aim to break-down barriers between biological disciplines and to repair broken links between modern "cutting edge" bioscience and its history. The pages receive numerous visits each day from sites world-wide, so there seem to be plenty of interested people out there.

Those just wanting the science should go directly from chapters 1 and 2 to Parts 2 and 3. For the non-scientist with a thirst for characters, history and politics, Part 4 alone might suffice. However, I imagine a typical reader will be someone who enjoyed *The Selfish Gene*, or Stephen Gould's *Ever Since Darwin*, and now seeks more fundamental answers. Such a reader should not have great difficulty with Parts 1, 2 and 4. Should the book run to further editions, these parts should not need major changes. Nevertheless, some readers will be challenged. After all, we are dealing with ideas which confounded even Darwin himself! Part 3 describes new information arising from the computational analysis of genomes. This is more specialized and rather speculative. Progress here is rapid and the account will need frequent updating.

The twentieth century was obsessed with genes. Because the genic viewpoint explained so much, it was easy to assume that it explained everything. The Victorian Romanes and the Victorian–Edwardian Bateson thought otherwise. We are just beginning to understand what they were trying to tell us. A major conclusion of the book is that, in preference to the genic theory, a modified "chromosomal" theory of speciation should receive more attention. Recent evidence on chromosomal speciation has been summarized by cytogeneticist Max King in *Species Evolution: The Role of Chromosome Change* (1993). Michael White, a leading advocate of the chromosomal theory, stated at the end of his *Modes of Speciation* (1978 ch.10, 324, 349):

> The "modern synthesis" of the 1940s is now 30 years old, and some attempts to revise it have not taken into account sufficiently the vast increases in knowledge that have occurred in recent years. The result is that ... students of biology are presented with a stereotyped dogma that leaves them with the impression that all the basic problems have been solved. The main aim of this book has been to ... point ... out how much still needs to be discovered before we can confidently construct the "new synthesis" of evolutionary theory some 25 or 30 years from now – a synthesis that may indeed be the *final* one [White's italics] in this field of knowledge. ... This book is hence only the forerunner of the definitive work on ... speciation that should be written, about the year 2000 ... .

If all goes well, the present book should appear "about the year 2000." However, I hasten to add that the "new synthesis" presented here is unlikely to be "final," or "definitive." I simply hope to move the boundaries of the unknown "a little further from our sphere of action."

PART ONE

Search for a Victorian

Figure 1.1
Charles Darwin (1809-82).

1 Evolution of Languages and Species

"Mary had a little lamb, its fleece was white as snow"
Old English nursery rhyme

This rhyme would readily be pronounced by most of the cast of *Pygmalion* (Shaw 1913), which included the "toff" Freddy Eynsford-Hill, a man of some social standing. However, the "pore flahr gel," Eliza Doolittle, would have said "Miree ader liawl laimb, sfloyce wors woyt ers snaa." Both sentences convey the *primary* information that Mary possessed an immature, white, sheep. In class-conscious England at the beginning of the twentieth century, the sentences would also have communicated the *secondary* information that Eliza was of "lower class" cockney origin, whereas other members of the cast were of "middle class" origin.

In modern genetic parlance, the secondary information constitutes a "reproductive barrier." Cockneys tend to marry cockneys and perpetuate cockney secondary information. The middle class tend to marry into the middle class and perpetuate middle class secondary information. Shaw's experiment (play) shows that by removing the difference in secondary information (a linguistic barrier) the reproductive barrier is also removed. We are told that Eliza lived happily ever after with the middle class Freddy. The metaphor here of language and its divergence into dialects will later serve to assist us on a more literal level as we revisit the question of the origin of biological species as articulated in 1859 by Charles Darwin (Figure 1.1).

The nursery rhyme does not tell us whether Mary's lamb was a Merino, or a Southdown, or some other *variety* of sheep. For many centuries varieties (races, breeds, strains) of animals and plants have been generated and maintained by selective breeding. Indeed, so important have been herds and crops to the survival of post hunter-and-gatherer human societies, that the crossing of different animals and plants must have been a subject of intense scrutiny for millennia. Since ancient times such observations have been recorded. *The Origin of Species by Means of Natural Selection* appears by far the most complete compilation and analysis of these observations, to which Darwin added many of his own. Even if the book had not contained Darwin's theory of natural selection, it would still have been a monumental work. Thomas Huxley (Figure 1.2) wrote (1860 ch.2, 78): "If they [Darwin's theoretical views] were disproved tomorrow, the book would still be the best of its kind, the most compendious statement of well-sifted facts bearing on the doctrine of species that has ever appeared."

A measure of this is that, more than a century later, we still go back to Darwin's work as a primary source. Such is the pace of research these days, that most modern works of science contain citations only to the literature of the preceding decade. When the subject is biological, however, many modern discussions begin with Darwin.

Figure 1.2
Thomas Huxley (1825-95).

PARALLELS BETWEEN LANGUAGES AND SPECIES

It is a matter of common observation that cats breed with cats to produce more cats, dogs breed with dogs to produce more dogs, and humans breed with humans to produce more humans. Thus, the simple idea arose that a species is a group of organisms whose members breed with each other, but not with members of other species. Members of one species are "reproductively isolated" from members of other species. But this was not always so. Once upon a time, some 80 million years ago, there were no cats or dogs or humans. There was some prototypic mammal which produced lines of offspring which came to differ from each other in various characteristics, including a special characteristic, the *inability* to cross successfully with other lines. Cats, dogs, and humans are linked genealogically "by descent" from a common ancestor. The evolution of languages provides a helpful metaphor for this.

Huxley noted in 1863 (ch.11, 458) that: "In the language that we speak in England, and in the language of the Greeks, there are identical verbal roots, or elements, entering into the composition of words. That fact remains *unintelligible* so long as we suppose English and Greek to be *independently* created tongues; but when it is shown that both languages are descended from *one* original, we give an explanation of that resemblance."

A book on the evolution of man by the geologist Charles Lyell included a chapter entitled "Origin and Development of Languages and Species Compared." He wrote (1863 ch.23, 467): "We may compare the persistence of languages, or the tendency of each generation to adopt without change the vocabulary of its predecessor, to the force of inheritance in the organic world, which causes the offspring to resemble its parents. The inventive power which coins new words or modifies old ones, and adapts them to new wants and conditions as often as they arise, answers to [resembles] the variety-making power in the animate creation [the world of biological organisms]."

Of the different dialects of a language (e.g., cockney) Lyell noted (ch.23, 466) that: "Between these dialects, which may be regarded as so many incipient languages, the competition is always keenest when they are most nearly allied." Darwin liked the parallel between genealogy and language and suggested (1859 ch.13, 422): "If we possessed a perfect pedigree of mankind, a genealogical arrangement of the races of man would afford the best classification of the various languages now spoken throughout the world; and if all extinct languages, and all intermediate and slowly changing dialects, had to be included, such an arrangement would, I think, be the only possible one." Thus, reproductive choices leading to the genealogical arrangement of races have been strongly influenced by linguistic barriers, so that classifications of races and languages closely match (Barbujani 1997).

LANGUAGE MORE THAN A METAPHOR

The parallel between the evolution of languages and the evolution of organisms took on a more profound significance in modern times when it was appreciated that DNA was a genetic code script (Watson and Crick 1953). Like the science of linguistics, biology could be considered a branch of a discipline which has come to be known as information science (Shannon 1948). Just as the language you are now reading consists of a linear sequence of letters, so DNA consists of a linear sequence of chemical "letters," called "bases." In DNA there are four bases (adenine, cytosine, guanine and thymine), which are abbreviated as A, C, G and T. DNA is the genetic material (the genotype) which is passed on from generation to generation. The DNA sequence of bases encodes another type of "letter," the "amino acids." There are twenty amino acids, with names such as aspartate, glycine, phenylalanine, serine and valine (which are abbreviated as Asp, Gly, Phe, Ser and Val). Under instructions received from DNA, amino acids are joined together in the same order as they are encoded in DNA, to form proteins. The latter, chains of amino acids which fold in complicated ways, play a major role in determining how we interact with our environment. In genetic parlance, the proteins determine our "phenotype." For example, in an organism of a particular species ("A") the twenty-one-base DNA sequence:

TTTTCATTAGTTGGAGATAAA,

read in sets of three bases ("codons"), conveys primary information for a seven amino acid protein fragment (PheSerLeuValGlyAspLys). All members of the species will tend to have the same DNA sequence, and differences between members of the species will tend to be rare and of minor degree. If the protein is fundamental to cell function it is likely that organisms of *another* species ("B") will have DNA which encodes the same protein fragment. However, when we examine their DNA we might find major differences compared with the DNA of the first species (the similarities are underlined):

T̲T̲CAGCCT̲CG̲TGG̲GG̲GA̲CA̲A̲G.

This sequence also encodes the above protein fragment, showing that the DNA contains the same *primary* information as in the first DNA sequence, but it is "spoken" with a different "accent." This "*secondary* information" might have some biological role. It is theoretically possible (but unlikely) that *all* the genes in an organism of species B would have this "accent," yet otherwise encode the *same* proteins. In this case, organisms of species A and B would be both anatomically and functionally (physiologically) *identical* while differing dramatically with respect to secondary information.

On the other hand, consider a single change in the sequence of species A to:

TTTTCATTAGTTGGAGT̲TAAA.

Here the difference (underlined) would change one of the seven amino acids. It is likely that such *minor* changes in a *very small* number of genes affecting development would be sufficient to cause anatomical and morphological differentiation *within* species A (e.g., compare a bulldog and a poodle, as "varieties" of dogs). In this case the secondary information would be hardly changed.

The view to be developed in this book is that, like the dropped Hs of Eliza Doolittle, the role of differences in secondary information is to *initiate*, and, for a while, *maintain*, reproductive isolation (Forsdyke 1996a). This can occur because the genetic code is a "redundant" or "degenerate" code; for example, the amino acid serine is not encoded by just one codon; there are six possible codons (TCT, TCC, TCA, TCG, AGT, AGC). In the first of the above DNA sequences (A) the amino acid serine (Ser) is encoded by TCA, whereas AGC is used in the second (B). On the other hand, the change in species A from GAT (first sequence) to GTT (third sequence) changes the encoded amino acid from aspartic acid (Asp) to valine (Val), and this could be sufficient to change the properties of the corresponding protein, and hence change the phenotype.

GEOGRAPHICAL DISTRIBUTION

Darwin's closest friend was the botanist Joseph Hooker (Figure 1.3), who travelled widely observing the distribution of plants. He noted (1860) that:

> The most prominent feature in distribution is that circumscription of the area of species, which so forcibly suggests the hypothesis that all the individuals of each species have sprung from a common parent, and have spread in various directions from it. ... With respect to this limitation in area, species do not essentially differ from varieties on the one hand, or from genera and higher [taxonomic] groups on the other; and indeed, in respect of distribution, they hold an exactly intermediate position between them, varieties being more restricted in locality than species, and these again [being] more [restricted] than genera. The universality of this feature (of groups having defined areas) affords to my mind all but conclusive evidence in favour of the hypothesis of similar forms having had but one parent or pair of parents. ... This circumscription of species and other groups in area, harmonizes well with that principle of divergence of form, which is opposed to the view that the same variety of species may have originated at different spots.

In seeking the origin of species, language serves as a powerful metaphor. The evolution of languages has a strong geographical isolation component. As far as we know languages have diverged from each other in a somewhat random manner, and people have usually learned the language spoken where they grew up. However, as international communication has increased (i.e., the geographical barrier has eroded) some governments, much concerned about linguistic purity, have realized that languages will inevitably blend unless active steps are taken to prevent this occurring. The geographical barrier has to be replaced by another barrier, some other form of isolation.

Figure 1.3
Joseph Hooker (1817-1911).

SUMMARY

The metaphor of language and its divergence into dialects will assist us on a more literal level as we revisit the question of the origin of biological species as articulated in 1859 by Charles Darwin. Biology is a branch of information science. DNA is a genetic code-script. The linear sequence of four bases in DNA, read in sets of three bases (codons), conveys information for the linear ordering of twenty amino acids in proteins. There are more codons than amino acids, and a particular amino acid may be specified by more than one codon. Thus, two DNAs may "speak" the same protein, but with different codon "accents."

2 Variation, Heredity, Phenotypic and Reproductive Selection

"The doctrine of progression, if considered in connection with the hypothesis of the origin of species being by variation, is by far the most profound of all that have agitated the schools of Natural History."

Joseph Hooker (1860)

Some aspects of the Romanes–Gulick version of the Darwin–Wallace theory of evolution by natural selection will now be outlined. Don't worry if you are not comfortable with it at first. We will clarify the details later. The present aim is to give you some feeling for where we are going, and introduce some general concepts, terms and definitions.

EVOLUTIONARY EVENTS

The theory incorporates four fundamental processes: variation, inheritance, phenotypic selection, and reproductive selection. Evolution results from the accumulation of many evolutionary "events," each involving some or all of these processes. Each process is itself composed of a series of sub-events, or sequential steps, but it is not essential to know what these are. Thus each process can be treated as a "black box" and manipulated as an entity, or object (without knowing what is inside the box). An evolutionary event begins with the process of variation and is followed by the process of inheritance continuing through the generations. Selection may follow variation.

The words "selection," "isolation," and "segregation" have led to much confusion. The words have the *same* meaning, although one may be preferred in a particular context. Selection is something which can be done when there is more than one of something. If there is only one of something, it is *already* selected (or isolated or segregated). In much of the literature, "selection" when used alone often means phenotypic selection; "isolation" when used alone often means reproductive selection.

(i) *Variation* (like does not produce like). There is variation in a biological species and the resulting variant members of the species are candidates for selection. Variation is a *process* which produces the variant
features of variant organisms. These variant features are themselves often
referred to as variations. Thus the word "variation" must be understood
in context either as the process, or as the result of that process. In the context of domestic breeding (selection for mating by man), Darwin (1859
ch.1, 41) noted that: "... as variations manifestly useful or pleasing to man
appear only occasionally, the chance of their appearance will be much
increased by a large number of individuals being kept; and hence this
comes to be of the highest importance to success. On this principle
Marshall has remarked, with respect to the sheep of parts of Yorkshire,
that 'as they generally belong to poor people, and are mostly in *small lots*
[Darwin's italics] they never can be improved.'"

Thus, from the outset, it was appreciated that *population size* (large or
small lots) could affect outcomes, and that variation had a *random* quality, at that time largely out of man's control. All the breeder could do was
select when variants appeared, and then appropriately mate. The breeder
could only select some characteristic which was observable (the "phenotype"). A variation (variant feature) not observable by man (or not
detected by some aspect of the natural environment in the case of natural breeding) would be ineligible as a candidate for phenotypic selection.
An initially undetectable variation, however, might cumulate in some
way with other variations to become detectable.

Hooker (1860) stressed the autonomous nature of variation ("creation by
variation") and its independence of environmental influences (1862a, 130):

> I have come to look upon the law of variation as I do on gravitation: local
> circumstances may mask its affects, but upon *itself* they cannot act
> [Hooker's italics]. ... A certain amount of centrifugal force of variation is
> distributed in certain proportions amongst the 12 peas in the pod; and
> except to arrest or retard the progress and amount of development of the
> individuals and their organs ... local circumstances are powerless. Give
> more nitrogen to pea number 1 and you will have more and greener
> leaves; but [the offspring of] its seeds will not be as green ... [if] they too
> are not supplied with nitrogen.

It should be noted here that the environment (provision of nitrogen) has
produced observable variants (plants with greener leaves), but these *non-
inherited* variant features (variations) reflect a different process from the
process Darwin and Hooker had in mind.

(ii) *Inheritance* (like produces like). There is inheritance of a newly
acquired variation by at least some of the offspring of variant members,

and inheritance of that variation, in turn, by some of their offspring, and some of the subsequent offspring. If a variation is not inherited (overtly or cryptically) then it may be the result of some environmental influence and is not strictly related to the evolutionary event under consideration.

(iii) *Phenotypic selection.* There is positive selection of members of the species with advantageous variations, and negative selection of members of the species with disadvantageous variations. These processes are "checks to indiscriminate variation" (Hooker 1860). Isolation on the basis of phenotypic adaptations (positive or negative) is the type of selection with which Darwin was primarily concerned. In this respect he and others spoke of "survival of the fittest" and the "struggle for existence." He used the term "*natural* selection," to distinguish this form of phenotypic selection from the *artificial* selection carried out by the domestic breeder. However, reproductive selection (see below) can *also* be natural or artificial.

(iv) *Reproductive selection.* The above three processes can operate so that advantageous phenotypic characteristics would tend to increase in a population, and disadvantageous phenotypic characteristics would tend to decrease in a population. They provide for the *adaptation* of species, and operate primarily at the level of the individual member of the species (individual selection). However, for the *origin* of species (i.e., a *divergence* between two population groups), a fourth process, reproductive selection (usually referred to as "*reproductive isolation*") is needed. To originate species, *groups* have to be selected. Speciation is a form of *group selection.* A group is a set of units. For our purposes, if there is only one unit in a set it is an individual, and does not constitute a group. A group whose members reproduce only asexually are, by definition, already reproductively isolated from other groups. Darwin noted (1859 ch.1, 42):

> Facility in preventing crosses is an important element of success in the formation of new races, at least, in a country which is already stocked with other races. In this respect enclosure of the land [geographical barrier] plays a part ... . On the other hand, cats, from their nocturnal rambling habits, cannot be matched [reproductively isolated], and, although so much valued by women and children, we hardly ever see a distinct breed kept up; such breeds as we do sometimes see are almost always imported from some other country, often from islands [geographical barrier].

An example of species *adaptation* alone (processes (i)–(iii)) is shown in Figure 2.1. A hypothetical prototypic giraffe/horse population of varied neck length (phenotype) is represented as a bell-shaped distribution when the number of animals (organisms) in small height ranges are plotted against overall height (Figure 2.1a). A severe drought removes food so

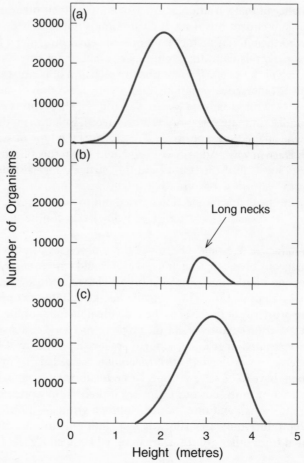

Figure 2.1
Species adaptation (serial, "monotypic," evolution) in response to phenotypic ("natural") selection.

we may imagine that *only* prototypic giraffe/horses which can stretch to nibble leaves on the highest branches of trees survive. At a subsequent point in time, the short-term selective factor (drought) is removed, but now the distribution is as shown in Figure 2.1b. The range of individuals with whom to mate has decreased. With time there will be a new bell-shaped distribution as shown in Figure 2.1c. In this example, there is *no alternative group* of organisms with which members of the surviving group can breed. Since the survivors are reproductively isolated by the operations of processes (i) to (iii), there is *no need* for additional reproductive isolation, as in process (iv).

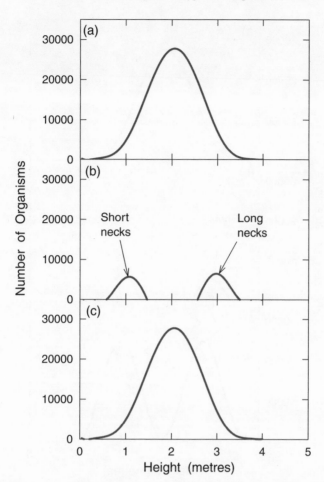

Figure 2.2
"Swamping" when other members of the parental group survive, but reproduction selection/isolation is absent.

However, Figure 2.2 shows another scenario, where the prototypic giraffe/horses with the longest necks survive (for the same reason as in Figure 2.1), and the prototypic giraffe/horses with the shortest necks survive, since (say) they are more readily able to migrate to a region where food is more abundant (Fig. 2.2b). With the removal of the short-term selective factor (drought) the two surviving groups merge and their members mate; over the generations the separate grouping is lost (Fig. 2.2c). If something essential for process (iv), reproductive isolation, had been present, then the adaptations might have been preserved. There would then have been a divergence between groups, perhaps into something horse-like

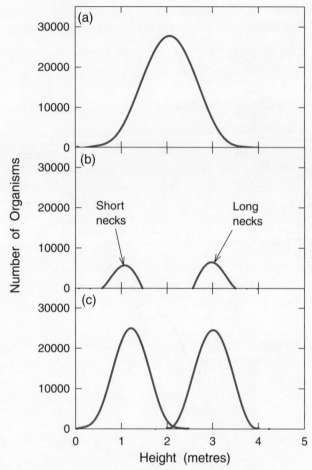

Figure 2.3
Species divergence (branching, "polytypic," evolution) when there is both reproductive and phenotypic selection/isolation.

and something giraffe-like. There would have been an *origin* of species, as shown in Figure 2.3. Even if the height ranges (phenotypes) of the two emerging groups overlap, as in Figure 2.3c, they are reproductively isolated.

This book is about the "something essential for process (iv)." In some cases the "something" may be a special case. For example, coinciding with the above drought there might have been a geological disturbance such that a canyon came permanently to separate the protohorse and protogiraffe populations (so that they were as if on separate islands). Over the generations operation of variation, inheritance, and phenotypic selection (processes (i) to (iii)) might further adapt members of these species in the direction of horse and giraffe, respectively. If a subsequent geological disturbance then removed the canyon the two groups would merge. However, by this time anatomical changes would have greatly impeded copulation between members of the groups. Thus reproductive isolation initially maintained by geographic barriers would be followed by reproductive isolation maintained by anatomical barriers.

The point here is that the formation of new species requires reproductive isolation. *Any* shape or form of reproductive isolation will suffice. The latter example is trivial, and unlikely to correspond to most evolutionary events. We seek a fundamental form of *initial* reproductive isolation which has operated in the *general* case of the majority of species, both past and present. In short, we seek *the* origin of species. We are not here concerned with the origin of the higher taxonomic divisions (genera, families, orders, classes) which would follow speciation events.

THE CYCLE OF THE GENERATIONS

As human beings we tend to perceive the end product of an act of love as a child, which grows into an adult. We think of the meeting of this adult with the "right" adult of the opposite sex as a new beginning, which can lead to another act of love and a further child, which grows into an adult. This can be regarded as a never-ending cyclic process, the cycle being reinitiated each time true lovers meet, and ending with the production of a mature adult.

Yet, *biologically*, copulation is far from completed when a new child is produced to grow into an adult. When the gamete of the father (spermatozoa) merges with the gamete of the mother (ovum) to produce a new cell (zygote), *two* genomes come to *co-exist* within a *single* individual (hybrid). The two distinct genomes *cooperate* to allow the development of the fertile egg into a multicellular adult organism with its own gonads. It is in the gonad of this offspring that the initial parental act of love is truly consummated. Here, in the process known as meiosis, the two parental genomes come together in a most intimate way to produce gametes with potential to give rise to the next generation. Thus, in biological terms, although the cycle of the generations can still best be considered as beginning with the meeting of adults "hit with Cupid's arrow," its termination does not occur until the ultimate consummation in the gonad of the offspring some time prior to an act of love in the next generation.

Within a species the cycle operates continuously. Different "varieties" within a species become distinguished from each other as new "species" when members of one variety are no longer able to reproduce with members of the other. This reproductive isolation is achieved when the cycle is interrupted at some stage. The problem of the origin of species is essentially that of determining what form that interruption *first* takes, and at what stage in the cycle of the generations it *usually* occurs. We will be discussing in chapter 4 the three fundamental barriers to cycle operation, one operating "pre-zygotically" to prevent gametes meeting and two operating "post-zygotically" after the gametes have merged. The Montagues and the Capulets had only the pre-zygotic barrier to keep apart Romeo and Juliet. The "hand of Nature" also has the two post-zygotic barriers. Figure 2.4 shows a single species diverging into two separate species and the time at which different barriers operate, *in the general case*. The barrier most proximal to the initial divergence is "hybrid sterility." A later barrier is "hybrid inviability," and an even later barrier is "pre-zygotic isolation." Thus, barriers tend to replace each other consecutively.

The fact of replacement has important consequences. Your dog may be tethered in your garden by a leash. But if you build a high fence, the leash is no longer necessary. If the initial barrier (the leash) is damaged or lost, it may not be noticed. The second barrier (fence) should suffice. On the other hand, the leash then becomes available for some other function. Thus,

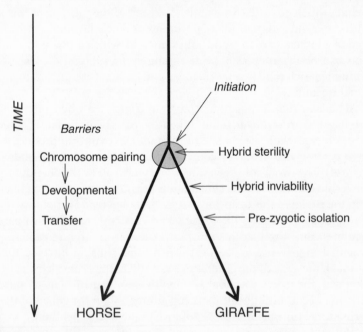

Figure 2.4
Consecutive appearance of different barriers mediating reproductive isolation, in the general case. At the point of divergence there is the chromosomal pairing barrier (hybrid sterility). This may be followed by a developmental barrier (hybrid inviability), and a gamete transfer barrier (pre-zygotic isolation).

following establishment of a second barrier, a first barrier may degenerate or change in a random way, or may find other employment. If a Sherlock Holmes then tried to discern whether there had been an earlier barrier than the fence, and what form it had taken, there might be a problem.

"BREEDING TRUE" FROM HYBRIDS

The word "hybrid" has to be used with care and understood in context. For example, African pygmies are usually less than five feet in height, whereas Europeans are usually greater than five feet. If a pygmy breeds with a pygmy the child normally grows to pygmy height. When a European breeds with a European, the child normally grows to European height. Thus, the height character can be said to "breed true" in these two racial groups. All the children of a pygmy couple grow to be small. All the children of a European couple grow to be tall (relative to a pygmy). With respect to height the children are not hybrids. However, if an African pygmy were to breed with an European, the child would be a hybrid with respect to height. The definition as a hybrid holds irrespective of the height the child actually attains as an adult. It is a description of genetic background, not of physical appearance.

Two pygmy individuals breeding together are likely to differ in numerous phenotypic characteristics besides height. With respect to these differing characteristics their children would be hybrids. Indeed, a strict definition requires that the children of all interbreeding individuals be considered as hybrids. Any offspring is the hybrid of its parents if the parents differ genetically (genomically) in some way, whether or not this genomic difference (difference in any part of DNA) correlates with a phenotypic difference. The only possible human offspring not a hybrid would be a female who inherited absolutely identical DNA sequences from both parents (an almost impossible event). All human males are hybrids since, even if the autosomes were identical, the sex chromosomes would differ (see chapter 19). The ancient Greek word for hybrid means an insult or outrage, with special reference to lust. At that time most hybrids were perceived as outrages on nature, as mongrels.

The problem of breeding true has long been of concern to animal and plant breeders, who have sought to obtain organisms with new features by crossing existing organisms with differing characters. They seek "constant hybrids." It is, of course, not feasible experimentally to cross pygmies with Europeans. Yet, since they are races within a species, there is no reproductive isolation (see chapter 3) and the offspring should be viable. Would these hybrids be short or tall, like the parents, or of some intermediate size? Although unlikely, the hybrids might be even taller than the Europeans? If the breeder sought, and successfully obtained, adults of (say) intermediate size, then the next step would be to cross a male and female from among these and observe the heights of the next generation. Would the intermediate size characteristic breed true (i.e., would the offspring be "constant hybrids"), or would there be a reversion to one or other adult forms?

Of course, these were the experiments carried out by Gregor Mendel with peas as described in elementary biology texts (Mendel 1865; Forsdyke 2001b). In the case of peas the height character appeared to depend on a single pair of allelic genes, with tallness being dominant over smallness, so that the first generation of hybrid plants were all tall (see chapter 18). When these first generation hybrids were crossed with each other, there was a reversion to the parental characters in the next generation according to the famous Mendelian ratios of three tall plants to one small plant (on average). The small plants were all homozygous for the small allele and so bred true with each other (like the corresponding small parental type of the original cross). Of the three tall plants, only one corresponded to the original tall parental type. The other two could be shown by further breeding studies still to be hybrids carrying a dominant allele for tallness and a recessive allele for smallness (Figure 2.5). Thus, no "constant hybrids" had been created as a result of the original cross. Indeed Hooker noted in 1860: "A very able and careful experimenter, M. Naudin, performed a series of experiments at the Jardin des Plantes at Paris, in order to discover the duration of the progeny of fertile hybrids. He concludes that the fertile posterity of hybrids disappears, to give place to the pure typical form of one or other parent."

Even if novel plants were obtained which did breed true (producing their like uniformly among all offspring from generation to generation), it would be most improbable that these constant hybrids would be reproductively isolated. In order to preserve the variant type, the breeder would have to be the isolator, taking precautions to ensure that there was no chance fertilization by parental pollen. From time to time reproductively isolated constant hybrids are reported, but these turn out to be special cases (e.g., polyploidy; see chapter 7).

An early student of hybridization in plants, Joseph Kölreuter (1733-1806), noted that the nature of the offspring was often independent of the parental origin of the characters (Roberts 1929 ch.2, 56-7). Tt and tT ($T_p t_M$, and $t_p T_M$) are equal hybrids (Figure 2.5). Thus, Max Wichura (1817-1866) observed that the genetic contributions of the father and mother were likely to be equal, despite the maternal gamete usually being larger (Roberts ch.6, 181-2). Carl von Nägeli (1817-1891), a botany professor in Munich, deduced that if a material substance were responsible for the transmission of characters between generations, then the substance (which he called "idioplasm") should be quantitatively similar in paternal and maternal gametes and thus would constitute a relatively small part of the maternal gamete (Roberts ch.6, 197-204).

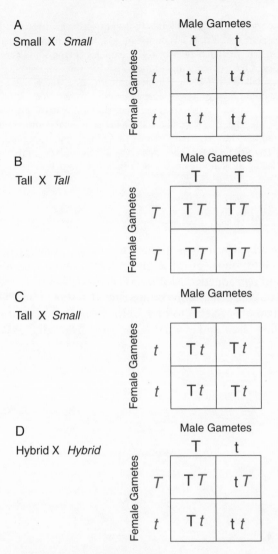

Figure 2.5
Distribution of allelic genes for tallness (T) and smallness (t) according to Mendel. A
diploid organism inherits one allelic gene for a particular character (e.g., height) from
its father, and one allelic gene from its mother. The paternal and maternal genes are
designated "allelic" because they are potential alternative versions ("alleles") and occupy corresponding positions on homologous chromosomes. The diploid state can be
written as TT for a homozygous tall organism and tt for a homozygous small organism. From these would be derived two types of haploid gametes (with either the
paternally- or maternally-derived character). Thus, the homozygous parents from
which gametes are derived might be designated $T_P T_M$, and $t_P t_M$, but for simplicity the
subscripts are omitted. For crosses between male and female homozygotes of the same
kind, the parental type offspring are produced (A and B). For crosses between male

and female homozygotes of different kinds (C), all offspring are hybrids with respect to the height character (i.e., Tt). If T is dominant to t, then all hybrids appear tall. If a male hybrid is crossed with a female hybrid (D), gametes of type T and t are produced in equal quantities by both parents, and of four offspring, three appear tall (TT, Tt, tT) and one appears small (tt). If T had not been dominant, some interaction between the T and t genomes might have produced a new phenotypic character, such as intermediate height. This character would not have "bred true," since all offspring would not have had the character (i.e., two would have still had the original homozygous parental characters, TT and tt). Thus "constant hybrids" would not have been produced. Note that Tt and tT ("reciprocal crosses") are equal hybrids; the order simply relates to the convention of placing the genome of paternal origin first, so that these might be designated as $T_P t_M$, and $t_P T_M$.

SUMMARY

The chapter seeks to clarify evolutionary terminology relating to the four fundamental evolutionary processes of variation, inheritance, phenotypic selection, and reproductive selection (isolation). The latter is achieved when the cycle of the generations is interrupted at some stage. The problem of the origin of species is to find, in the general case, what form the interruption first takes, and at what stage it usually occurs.

3 Darwin's Difficulties

> "But to the day of my death I will always maintain that you have been too sharp-sighted on hybridism; and the chapter on the subject in my book I should like you to read: not that, as I fear, it will produce any good effect, and be hanged to you."
>
> A letter from Darwin to Huxley
> (1868; Darwin and Seward 1903 ch.4, 287)

Ayala and Fitch (1997) note that: "The most serious difficulty facing Darwin's evolutionary theory was the lack of an adequate theory of inheritance that would account for the *preservation* through the generations of the variations on which natural selection was supposed to act." Darwin's Victorian critics were very concerned about the preservation of variation. Since we wish as much as possible to let the Victorians tell the story, we must pause, read, and reread, the sometimes convoluted prose of one who was both Darwin's most staunch defender and most penetrating critic, Thomas Henry Huxley (1888 ch.10, 290). The effort is usually well rewarded:

It is not essential for Mr. Darwin's theory that anything more should be assumed than the facts of heredity, variation, and unlimited multiplication; and the validity of the deductive reasoning as to the effects of the last (that is, of the struggle for existence which it involves) upon the varieties resulting from the operation of the former [variation and heredity]. Nor is it essential that one should take up any particular position in regard to the mode of variation, whether, for example, it takes place *per saltum* [in a punctuated fashion] or gradually; whether it is definite [perceptible] in character, or indefinite [imperceptible]. Still less are those who accept the theory bound to any particular views as to the causes of heredity and variation.

Thus, the fact that organisms varied, and that their offspring often carried similar variations (i.e., there was some genetic component to the variation), was not in dispute. Darwin's primary point was that variations might confer some selective advantage, which might increase the reproductive success of a variant organism relative to organisms which had not so varied. In Huxley's words (1860 ch.2, 71):

The Darwinian hypothesis ... [is that] all species have been produced by the development of varieties from common stocks; by the conversion of these, first into permanent races and then into new species, by the process of *natural selection*, which process is essentially identical with that artificial selection by which man has originated the races of domestic animals, the *struggle for existence* taking the place of man, and exerting, in the case of natural selection, that selective action which he performs in artificial selection [all italics are Huxley's].

But man *both* selects (phenotypic isolation) *and* matches (reproductive isolation). An engineering professor (Jenkin 1867) reminded Darwin of a problem first perceived in 1842 (Vorzimmer 1963), that under natural conditions there would be a "swamping" effect due to the blending of parental characters in the offspring (Figure 2.2). The "struggle for existence," as implied by Darwin (phenotypic selection), *itself* only produces reproductive isolation in the sense of causing the deaths of non-survivors who are thus no longer available for mating (Figure 2.1). In his above remark Huxley seems to be discounting the role of reproductive isolation leading to the *divergence* of members of an incipient species away from the still-surviving members of the parental group. However, perhaps dimly perceived, this was actually his major concern.

HUXLEY'S DIFFICULTY

Darwin *based* the theory of natural selection on the premise that observations on anatomical variants arising under domestication ("morphological species") were highly relevant to the process by which natural ("physiological") species had arisen. If that key assumption were false, then .... But let Huxley (1863 ch.11, 463) explain: "Every hypothesis is bound to explain, or, at any rate, not be inconsistent with, the whole of the facts which it professes to account for; and if there is a single one of these facts which can be shown to be inconsistent with (I do not mean inexplicable by, but contrary to) the hypothesis, the hypothesis falls to the ground."

The most serious difficulty in Huxley's view was that crosses between members of anatomically different "species" arising under domestication are usually fertile; however, if crosses ("first crosses") between members of different *true* species (e.g., horse and ass) can be brought about, and if offspring are produced (e.g., a mule), then that offspring (hybrid) is usually sterile (i.e., "second crosses" fail). Thus races of domestic animals differ in a quite fundamental manner from true biological species. Offspring of crosses between the former are fertile, whereas offspring of crosses between the latter are not. In an 1863 letter to the author Charles Kingsley, Huxley pointed out (Huxley 1900 vol.1, ch.17, 239) that Darwin: "... has shown that selective breeding is a *vera causa* for [true cause of] morphological [domestic] species; but he has not yet shown it a *vera causa* for physiological [natural] species."

In his *Six Lectures to Working Men* (1863 ch.11, 464) Huxley added:

At present, so far as experiments have gone, it has not been possible to produce this complete physiological divergence by selective breeding. ... If it could be proved, not only that this *has* not been done, but that it *cannot* be done; if it could be demonstrated that it is impossible to breed selectively, from any stock, a form which shall not breed with another, produced from the same stock; and if we were shown that this must be the necessary and inevitable results of all experiments, I hold that Mr. Darwin's hypothesis would be utterly shattered [all italics are Huxley's].

Thus, Huxley argued that if you are going to extrapolate observations in one system (artificial) to a second system (natural), then the two systems should be similar in as many aspects as possible, and certainly in aspects as fundamental as the sterility of hybrids.

In *The Variation of Animals and Plants under Domestication* (1875 ch.19, 162), Darwin acknowledged this as a major stumbling block: "The fact remains indisputable that [members of] domestic varieties of animals and of plants, which differ greatly from one another in structure, ... are extremely fertile when crossed [with members of allied varieties]; and this seems to make a broad and *impassable* barrier between domestic variety and natural species."

He had attempted to rescue his hypothesis (Darwin 1859 ch.8, 272) by suggesting that the difference was one of degree, and that sterility of species hybrids: "is not a special endowment, but is *incidental* on *slowly acquired modifications* ... in the *reproductive systems* of the forms which are crossed." Hence, in Darwin's view *first* functional morphological differences would appear, providing a basis for natural selection, and *then* reproductive isolation would be furthered by the development of a hybridization barrier. In the case of species arising under domestication, the period of domestication may have been too short for the hybridization barrier to develop. Huxley in his 1863 letter was quite prepared to accept this (Huxley 1900 vol.1, ch.17, 239): "I entertain little doubt that a carefully devised system of experimentation would produce physiological [i.e., reproductively isolated] species by selection, only the feat has not been performed yet." Nor had the feat been performed sixty years later. This led the Cambridge biologist William Bateson to write (1925):

> Nothing that has happened since [Huxley's remarks] at all mitigates the seriousness of this [Huxley's] criticism. ... I doubt whether many of those best acquainted with modern genetics are so sanguine as Huxley was, that by the most carefully devised system of experimentation are we in the least likely to produce physiological species by selection. Rather we have come to suspect that no amount of selection or accumulation of such variations as we commonly see contemporaneously occurring can ever culminate in the production of that "complete physiological [reproductive] divergence" to which the term species is critically applicable.

HOOKER'S DIFFICULTIES

Hooker (1860) thought Darwin had put too much emphasis on the power of natural selection, and had paid insufficient attention to the primacy of variation: "It is to variation that we must look as the means which Nature has adopted for peopling the globe with those diverse existing forms which, when they tend to transmit their characters unchanged through many generations, are called species." In a letter to Darwin, Hooker wrote (1862b, 570):

> You must remember that it is neither crossing nor Natural Selection that has *made* so many divergent human individuals, but simply *Variation* [Hooker's bold emphases]. Natural Selection, no doubt, has hastened the process, intensified it (so to speak), has regulated the lines, places, etc., etc., etc., in which, and to which, the races have run and led, and the number of each, and so forth; but, given a pair of individuals with power to propagate, and [an] infinite [time] span to procreate in, so that not one be lost, or that, in short, Natural Selection is not called on to play a part at all, and I maintain that after n generations you will have extreme individuals as totally unlike one another as if Natural Selection had extinguished half. If once you hold that Natural Selection can make a difference, i.e., create a character, your whole doctrine tumbles to the ground. Natural Selection is as powerless as physical causes to make a variation; the law that "like shall not produce like" is at the bottom of [it] all, and is as inscrutable as life itself.

SPECIES DEFINED MORPHOLOGICALLY OR REPRODUCTIVELY?

Much of the argument surrounding Darwin's theory rests on an acceptable general definition of biological species, a matter of some controversy. Lyell (1863 ch.10, 389) noted that: "From the time of Linnaeus ... a species

... [has been held to] consist of individuals all resembling each other, and reproducing their like by generation." This is equivalent to the most secure modern definition of a biological species as a group of organisms which is *reproductively isolated* from other groups of organisms (Templeton 1989). For a sexually reproducing group, members of the group are infertile with all other organisms, except those organisms which belong to the group. Along similar lines, Vulic and co-workers (1997) define species: "as a population of organisms capable of sharing their gene pool through mating and genetic recombination."

Those concerned with classification of organisms ("systematists") tended to use anatomical criteria, rather than physiological (including reproductive) criteria. However, Darwin (1859 ch. 2, 258) held that the capacity of members of two different species to cross: "is often completely independent of their systematic affinity, or of any recognizable difference in their whole organization. On the other hand, these cases clearly show that the capacity for crossing is connected with constitutional differences imperceptible to us, and confined to the reproductive system."

Huxley agreed, noting (1863 ch.11, 428): "The evidence ... is against the argument as to *any* limit to divergences, so far as structure is concerned; and in favour of a *physiological* limitation. By selective breeding we [the breeder] can produce structural divergences as great as those of species, but we *cannot* produce equal physiological divergences." Distinction between anatomical ("external") and physiological ("internal") characters was also made by Lyell (1863 ch.21, 420): "None of the observations are more to the point, as bearing on the doctrine of what Hooker terms 'creation by variation', than the great extent to which the *internal* characters and properties of plants, or their physiological constitution, are capable of being modified, while they exhibit *externally* no departure from the normal form."

The need to include reproductive criteria when defining species was later emphasized by Bateson (1913 ch.11, 236):

All constructive theories of evolution have been built on the understanding that we know if the relation of varieties to species justifies the assumption that the one phenomenon [varieties] is a *phase* of the other [species formation], and that each species arises ... from another species either by one, or several, genetic steps. In the varieties we have accustomed ourselves to think that we see these steps .... [However], complete fertility of the results of intercrossing [of members of different "species"] is, and I think *must* rightly be regarded, as *inconsistent* with actual specific difference [definition as being distinct species].

Darwin (1875 ch.19, 163) had appeared to acknowledge this: "We can only escape the conclusion that [members of] some *species* are fully fertile when [inter]crossed [with members of other "species], by determining to designate [by defining] as *varieties* all the forms which are quite fertile. This high degree of fertility [between "species"] is, however, rare ... ." He is actually, as we shall see later, seeking to represent the distinction between varieties and species as a man-made problem arising from error in definition (1875 ch.19, 173): "With respect to varieties which have originated in a state of nature, it is almost hopeless to expect to prove by direct evidence that they have been rendered mutually sterile; for if even a trace of sterility could be detected, such varieties would at once be raised by almost every naturalist to the rank of distinct species."

HYBRID STERILITY IS "INCIDENTAL"

Darwin in 1859 (ch.8, 245) notes:

> The view generally entertained by naturalists is that [members of different] species, when intercrossed, have been *specially endowed* [divinely endowed] with the quality of sterility [in the offspring], *in order to prevent* the confusion of all organic forms. This view certainly seems *at first* [to superficial analysis] probable, for species within the same country could hardly have kept distinct had ... [their members] been capable of crossing freely. The importance of the fact that hybrids are very generally sterile, has, I think, been much underrated by some late writers.

Darwin in 1875 (ch.19, 170) gives a fuller account of his thinking: "In considering the probability of *natural selection* having come into action in rendering [members of different] species mutually sterile, one of the greatest difficulties will be found to be in the existence of *many graduated steps from slightly lessened fertility to absolute sterility.*"

Having adopted the view that evolutionary change can only be brought about by natural selection, Darwin is confronted with a logical impasse. How can natural selection contribute to reproductive isolation when natural (phenotypic) selection is, after all, about selection of *individuals* for reproductive *success*, not for reproductive *failure*? However, Darwin continues:

> It may be admitted ... that it *would* profit an incipient *species* [variety] if it [the offspring] were rendered in some slight degree sterile when [members of the species were] crossed with ... [members of the] parent-form [parent species] or with some other variety [different incipient species]; for thus fewer bastardized and deteriorated offspring would be produced to comingle their blood [blend genetic characteristics] with the new species in the process of formation [with members of the incipient species].

Thus, with respect to the *group* (species), hybrid sterility could be advantageous. However, with respect to the individual Darwin continues (170-1):

> Take the case of any two species [members of] which, when crossed, produce few and sterile offspring; now what is there which could favour the survival of those *individuals* which happened to be endowed in a slightly higher degree with mutual infertility, and which thus approached by one small step towards absolute sterility? Yet an advance of this kind, if the theory of natural selection be brought to bear, *must* have incessantly occurred with many species, for a multitude are mutually quite barren.

Hence, from the viewpoint of natural selection acting on *individuals*, infertility should be selectively disadvantageous with few offspring being produced to propagate the infertile trait. Darwin further notes (171): "An individual animal ... if rendered slightly sterile when crossed with [a member of] some *other* variety, would *not* thus *itself* gain any advantage or *indirectly* give any advantage to the other individuals of the *same* variety, thus leading to their preservation." He is here contradicting himself in that he has already admitted (see above) that: "... it *would* profit an incipient *species* [variety] if it [its members] were rendered ... sterile." This would indeed "indirectly give ... advantage to the *other individuals* of the *same* variety" [i.e., the collective of individuals which constitutes an incipient species].

It will be shown later (chapter 4) that selective fertility (infertility with one group, while retaining fertility with another) might not necessarily result in the production of fewer offspring when viewed from Darwin's natural selection perspective. He does not make clear that a small increase in sterility marginally increases the *renouncement* of citizenship of the parent group, and marginally *affirms* citizenship of the incipient group. Whether this turns out to be advantageous depends on the *future* struggle of group versus group (group selection). Whereas a person contemplating migration to a new country can often weigh in advance the advantages of a change in citizenship, this possibility is not open to an organism "contemplating" a change in species. There are no options. Random forces determine whether or not the organism will be exiled.

Thus, variations which either increase or decrease hybrid sterility, pressure an organism towards or away from an incipient group. If a line is *diverging*, then at any moment the question arises as to whether an individual belongs to the parent group or to the variant group (incipient species). At that moment the adaptive benefits of an individual belonging to (reproducing with members of) one group rather than another would depend on (i) the degrees of relatedness to members of the groups (i.e., does it, in some fundamental genetic sense, belong more to the parental group or to the varietal group), and (ii) the results of *future* selection of members of the groups. Natural selection is constantly acting on members of both groups, but the *assignment* of individuals to one group or the other is an arbitrary process, *independent of natural selection*. Hybrid sterility with respect to the parental group enhances the purity of a variant group so that its members can better avail themselves of the advantages (if any) of group membership. Thus hybrid sterility in an individual may "indirectly give ... advantage to other individuals of the same variety."

However, Darwin (1875 ch.19, 171) concludes:

As [members of different] species have *not* been rendered mutually infertile through the accumulated action of natural selection [a fact he believed he had demonstrated], and as we may safely conclude ... that they have not been endowed through an act of creation with this quality [here he seems to appeal to common sense], we must infer that it has arisen *incidentally* during their *slow* formation in connection with *other and unknown changes in their organization*.

While stressing the importance of purposeful, non-incidental (non-random), reproductive isolation for the successful *domestic* breeding of anatomical variants (a relatively rapid process), he postulates a slow incidental (i.e., random) emergence of reproductive isolation, an isolation resulting from "*other and unknown changes in their organization*," which would favour the successful breeding of advantageous anatomical variants arising under natural conditions. These days we would describe this (politely) as vague "hand waving," or (impolitely) as "smoke and mirrors." But, as we shall see in Part 2, unknowingly Darwin may have been quite near the mark.

INTERNAL FORCES

Huxley (1880 ch.7, 240) states that Darwin could "be trusted always to state the case against himself as strongly as possible"; yet, in places, Darwin appears quite dogmatic. He repeats, time and time again that: "the term species ... does not essentially differ from the term variety" (1859 ch.2, 52), "species are only strongly marked and permanent varieties" (1859 ch.2, 56), "neither sterility nor fertility affords any clear distinction between species and varieties" (1959 ch.8, 248), "there is no essential distinction between species and varieties" (1859 ch.8, 276), "there is no fundamental difference between species and varieties" (1859 ch.8, 278).

Repetition, they say, is necessary to teach the fools the rules, and the wise the lies. After so many years of study Darwin might, with justification, have thought some of his critics a little foolish. Indeed, he used to reproach Huxley for his "pernicious insistence on the need for experimental verification" (1891 letter to Romanes; Huxley 1900 vol.2, ch.17, 292). But some of Darwin's critics were not foolish. They included Mendel himself. With the exception of Huxley, the major concern of serious Victorian critics was the apparent inutility of the characters which distinguished varieties or closely allied species. It was agreed that natural [phenotypic] selection should work to favour advantageous variants and disfavour disadvantageous variants, but the observed differences between members of closely allied species often appeared without utility (i.e., were "neutral"), and so were unlikely targets for natural selection. This was entirely consistent with Hooker's "creation by variation" (see chapter 2), meaning that variants would appear spontaneously. Of Darwin's work Mendel is reported to have said (Iltis 1932 ch.8, 103): "This much already seems clear to me, that nature [natural selection] does not modify species in any such way, so some *other force* must be at work." In his famous 1865 paper Mendel elaborated on this "force" (Iltis 1932 ch.11, 158):

> If the change of environment of growth [phenotypic selective factors] were the sole cause of variability, we should expect that those domesticated plants which had been cultivated for many centuries under almost identical conditions would by now have recovered their constancy [stopped varying]. But this, as we know, is not the case, for it is among these plants that we find, not only the most diversified, but also the most variable forms. ... It seems more probable that, in this matter of the variability of domesticated plants, there has been at work a factor to which hitherto little attention has been paid.

From 1866 to 1873 Mendel corresponded with Nägeli, who is reported (Iltis 1932 ch.14, 188) to have: "rejected ... selection as an active factor in the origin of species, thus setting himself in strong opposition to Darwin. In his view the progressive change in living forms was to be explained as the expression of *a progressive tendency inherent in living matter*, a tendency which was *not* to be regarded mystically, but as dependent upon mechanical [chemical?] forces." Thus, Nägeli took Mendel's criticism one step further in suggesting (vague though it might be) an *internal* "mechanical" cause of the origin of species, independent of natural selection.

CREATIONISM

Darwin had some sense of the three major objective criticisms (hybrid sterility, swamping, inutility; Vorzimmer 1963). Contrary to common belief, it was probably his awareness of these criticisms, rather than a fear of confronting religious orthodoxy ("creationism"), that caused him to delay submission of his theory for publication. Alfred Wallace seems to have had fewer reservations (a position held throughout his life), and his paper in 1858 forced Darwin's hand. However, as quoted by Huxley (1863 ch.11, 448), Darwin's position regarding the creation of living organisms was quite clear: "Given the origin of organic matter, supposing its creation to have *already* taken place, my object is to show in consequence of what laws and what demonstrable properties of organic matter, and of its environments, such states of organic nature as those with which we are acquainted must have come about."

Darwin took the existence of at least one form of life as given. He wanted to know how, from that one form, further forms (species) would have arisen. Huxley (1871 ch.5, 165) also drew a line: "In the ultimate analysis everything is incomprehensible, and the whole object of science is to reduce the fundamental incomprehensibilities to the smallest possible number."

My only contribution to this topic is a letter in a Canadian newspaper following its publication of an interview with Dawkins. The editor gave the letter the title "Do you think God is in a hurry?" (Forsdyke 1994c):

When interviewed by Thomas Bass, Oxford zoologist Richard Dawkins tells how one reader of his book *The Selfish Gene* felt that the whole of his life had become empty and the universe no longer had a point (The Lives and Times of the Selfish Gene, Oct. 18). Neither in the interview nor in his book does Dr. Dawkins clarify where his science ends and religion begins.

Adopting the style of Dr. Dawkins, one can answer this by saying that it depends on whether you think "God" is in a hurry. If God is in a hurry, then He/She has had to work hard to create the universe together with you and me. However, if God is not in a hurry, then it is sufficient to sprinkle some matter and energy around and then come back three billion years later.

The evolutionary ideas of Darwin, Haldane, Hamilton and others, so ably expounded by Dr. Dawkins, tell us quite simply that, given enough time, life will self-assemble. Since theologians maintain that God is eternal, He/She cannot be in a hurry; so there is no fundamental disagreement between the God idea and evolution. Unfortunately, to prop up our belief in God, theologians have attributed qualities to God which science can disprove [better explain]. This does not mean that the God idea is invalid, only that theologians picked the wrong arguments. For example, theologians have for centuries used our sense of wonder at the complexity of, say, the human eye to posit that there must be some supreme creator. Dr. Dawkins points out that, given enough time, eyes will evolve. His last book, *The Blind Watchmaker*, expounds this very well.

Much the same point was made by Kingsley (Darwin and Seward 1903 ch.3, 225) in his 1862 story of: "A heathen Khan in Tartary who was visited by a pair of proselytising Moollahs. The first Moolah said: 'Oh Khan, worship my God. He is so wise that he made all things.' But the second Moolah won the day by pointing out that his God is 'so wise that he makes all things make themselves.'" However, Romanes (1895b ch.2, 78-9) wondered how such a God could be a benevolent "personal" God:

> Looking at the outcome, we find that more than one half the species which have survived the ceaseless struggle are parasitic in their habits, lower and insentiate forms of life feasting on higher and sentient forms; we find teeth and talons whetted for slaughter, hooks and suckers moulded for torment – everywhere a reign of terror, hunger, sickness, with oozing blood and quivering limbs, with gasping breath and eyes of innocence that dimly close in deaths of cruel torture! ... If ... we see a rabbit panting in the iron jaws of a spring trap, and in consequence abhor the devilish nature of the being who, with full powers of realizing what pain means, can deliberately employ his whole faculties of invention in contriving a thing so hideously cruel; what are we to think of a Being who, with higher faculties of thought and knowledge, and with an unlimited choice of means to secure His ends, has contrived untold thousands of mechanisms no less diabolical? As far as Nature can teach us ... it does appear that the scheme ... is the product of a Mind which differs from the more highly evolved type of human mind in that it is immensely more intellectual without being nearly so moral.

We should attend again to Huxley (1863 ch.11, 449) before leaving this topic: "All human enquiry must stop somewhere; all our knowledge and all our investigation cannot take us beyond the limits set by the finite and restricted character of our faculties, or destroy the endless unknown, which accompanies, like its shadow, the endless procession of phenomena."

"SO HOPELESSLY WRONG"

Huxley in an article "The Coming of Age of *The Origin of Species*" (1880 ch.7, 229) was cautious: "History warns us ... that it is the customary fate of new truths to begin as heresies and to end as superstitions; and, as matters now stand, it is hardly rash to anticipate that, in another twenty years, the new generation, educated under the influences of the present day, will be in danger of accepting the main doctrines of the '*Origin of Species*,' with as little reflection, and it may be with as little justification, as so many of our contemporaries, twenty years ago, rejected them."

At least one member of the new generation, Bateson (1913 ch.11, 248), was not superstitious:

> The many converging lines of evidence point so clearly to the central fact of the origin of the forms of life by *one evolutionary process* that we are compelled to accept this deduction; but as to almost all the essential features,

whether of cause or mode, by which specific diversity [the diversity of species] has become what we perceive it to be, we have to confess an ignorance nearly total. The transformation of masses of population by imperceptible steps guided by *selection*, is, as most now see, so inapplicable to the facts, whether of variation or specificity, that we can only marvel both at the want of penetration displayed by the advocates of such a proposition, and the forensic skill by which it was made to appear acceptable even for a time.

In a letter to Hooker in 1888 Huxley admitted that he had not found Darwin easy reading (Huxley 1900 vol.2, ch.12, 192): "I have been trying to set out the argument of the '*Origin of Species*,' and reading the book for the nth time for that purpose. It is one of the hardest books to understand thoroughly that I know of, and I suppose that is why even people like Romanes get [it] so hopelessly wrong." That Romanes may not have been quite "so hopelessly wrong" will become evident later.

Near the end of his life (1894) Huxley addressed the Royal Society (Huxley 1900 vol.2, ch.22, 390):

Those who wish to obtain some clear and definite solution of the great problems which Mr. Darwin was the first person to set before us, in later times must base themselves upon the facts which are stated in his great work, ... and must pursue their enquiries by the methods of which he was ... so brilliant an exemplar throughout the whole of his life. You must have his sagacity, his untiring search after the knowledge of fact, his readiness always to give up a preconceived opinion to that which was demonstrably true, before you can hope to carry his doctrines to their ultimate issue.

With complete confidence that you, dear reader, are well endowed with these qualities, in the next chapter we will consider some of the facts relating to the problem of hybrid sterility.

SUMMARY

Of three major criticisms of Darwin's theory (hybrid sterility, swamping, inutility), the first (Huxley's difficulty) is of overwhelming importance. In the context of Darwin's struggle to deal with Huxley's arguments, we discuss species definitions, mysterious "internal" forces controlling selective fertility, group selection (anathema to some evolutionists), and creationism. Huxley considered Romanes to have got it all "so hopelessly wrong."

4 Hybrid Sterility

"One trouble with Darwinism is that, as Jaques Monod perceptively remarked, everyone *thinks* he understands it" [Dawkin's italics].
Richard Dawkins (1986 Preface, xi)

The diversity of organisms is such that few generalizations are absolute in biology. The biological universe consists of a broad mass of organisms to which a given generalization may apply, as well as special cases to which the same generalization may not apply. The study of special cases can be illuminating ("Treasure your exceptions" was Bateson's admonition), but that they may be special cases, perhaps of little general relevance, should always be borne in mind. One of Darwin's great strengths was his ability to winnow the relevant from the irrelevant, but when considering sterility even he could be confused. For we now know that there are many types of sterility. Bateson noted (1913 ch.11, 238):

> Breeders know that sterile animals and plants occasionally appear ... , but it is more in accordance with probability that the sterility in these sporadic instances should be regarded as due to [a] defect [in some Mendelian character] than that it should be thought comparable with that of sterile hybrids. ... The distinction between ... several kinds of sterility was ... not understood in Darwin's time. The comparison, for example, which he [Darwin] instituted between the [sporadic] sterility of ... [plants with withered] anthers and that of hybrids no longer holds, for at least in those cases in which the nature of ... [withered] anthers have been genetically investigated ... they proved to be a simple recessive character. Nor can we easily suppose that the attempt made by Darwin to suggest resemblance between the sterility produced by *unnatural conditions* and that of hybrids has any physiological justification.

Whereas Darwin tended to blur the lines between various types of sterility, there is in Bateson's remark a distinction between (i) a sterility due to "unnatural conditions" (ii), a sporadic sterility due to changes in genes contributing to discrete functions required for crossing between individuals (e.g., anthers), and (iii) the phenomenon of hybrid sterility.

PROPITIOUS ENVIRONMENTAL CONDITIONS

Reproduction is sensitive to environmental cues. Most plants cross on a seasonal basis. Thus the cycle of the generations can be interrupted for a period until environmental conditions are propitious for its continuance. Following fertilization many plant zygotes remain dormant as seeds, which will develop months later when winter is over. Not having this option, many animals pair only when it is propitious to do so. Mary's little lamb was almost certainly a spring lamb, reflecting the fact that the parents had been "on heat" in response to changes in hormone levels at a discrete time of the year.

Even the unicellular yeast has mating types (sexes). When nutrients are abundant yeast cells reproduce by relatively simple (mitotic) division. When nutrients are less abundant, yeast cells mate and undergo more complicated (meiotic) divisions before forming spores which may remain dormant until nutrients become more abundant. In yeast, as in many other micro-organisms, the mating is triggered by adverse conditions, implying that the mating is of adaptive value in such conditions. Many organisms considered higher up the evolutionary scale have no dormancy option, and so crossing occurs only when the environment is likely to be propitious for the survival of offspring (e.g., when nutrients are abundant). That such an arrangement would have arisen by classical Darwinian natural selection is quite obvious. Organisms which produced offspring when it was not propitious to do so would not have left many ancestors to propagate the trait.

Quite plausibly, Darwin (1859 ch.1, 8-9) ascribes to "unnatural conditions" the fact that animals of the *same* species brought from the wild into captivity often do not reproduce. Such conditions may be sensed as unlikely to be propitious for the production of offspring: "Nothing is more easy than to tame an animal, and few things more difficult than to get it to breed freely under confinement, even in the many cases when the male and female unite."

However, he suggests (1859 ch.8, 265-6) that the class of unnatural conditions includes the attempt to cross (form hybrids) between members of *different* species:

Thus we see that when organic beings are placed under new and *unnatural* conditions, and when hybrids are produced by the *unnatural* crossing of [members of] two species, the reproductive system, independently of the general state of health, is affected by sterility in a very similar manner. In one case the conditions of life have been disturbed, though often in so slight a degree as to be inappreciable by us; in the ... case ... of hybrids, the external conditions have remained the same, but the organization has been disturbed by two different structures and constitutions being blended into one. For it is scarcely possible that two organizations should be compounded into one, without some disturbance occurring in the development ... .

He here recognises that species, as opposed to varieties, have crossed some threshold such that the "organizations" of their members, will not permit the compounding "into one" with a member of another species, "without some disturbance." However, using the blanket term "unnatural conditions" he links phenomena which are quite disparate. There is a fundamental distinction between these two kinds of "unnatural conditions." In the case of members of the same species, normally fertile individuals may become infertile temporarily during captivity (probably because of some stress-induced hormonal disturbance). However, the infertility between members of different species (hybrid sterility) is protracted. Darwin noted (1859 ch.8, 266) that even in the few cases where: "hybrids are able to breed *inter se* [i.e., some offspring are produced], they transmit to their offspring from generation to generation the same compounded [impaired] organization, and hence we need not be surprised that their sterility, though in some degree variable, rarely diminishes."

Even more important, the sterility of the hybrid offspring of crosses between members of different species is accompanied by a most marked morphological change, gonadal degeneration (atrophy). There is usually no such gonadal degeneration in any offspring which may arise from crosses between members of the same species held in captivity. Furthermore, pair selection in captivity is often made by the capturer, rather than by the captured. There are hints that, in the wild, animals can detect cues as to the likelihood of reproductive success with a particular partner, and this may influence mate choice (see chapter 8).

Thus, it seems quite appropriate for Bateson (1913 ch.11, 239) to have concluded: "Nor can we easily suppose that the attempt made by Darwin to suggest resemblance between the sterility produced by unnatural conditions and that of hybrids has any physiological justification." Indeed, Bateson (1913 ch.11, 247) strongly doubted the general validity of observations on sterility produced by unnatural conditions: "Nothing short of the most familiar acquaintance with the habitual behaviour of individuals, and of certain strains kept under constant scrutiny for several years would allow the experimenter to form reliable judgements as [to] the value to be attached to observations of this class."

GONADAL DEGENERATION

Unlike the cells of other organs, some cells of animal gonads (testis, ovary) will become gametes (spermatozoa, ova). The DNA content of the latter (haploid) is half that of other body cells (diploid). Diploid cells contain two haploid sets of chromosomes, one derived from the organism's male parent and one from the organism's female parent. Cell divisions occur in most tissues during development and in adult life. Division occurs after DNA has replicated to double the total DNA content of the parental cell, and the daughter cells contain the same amount of DNA as in parental cells prior to the replication of their DNA. This process of *mitotic* cell division, as might occur in a typical diploid cell, is shown in Figure 4.1a.

On the other hand, to produce haploid gametes, germ cells undergo a special form of cell division (meiosis). This includes the pairing of homologous chromosomes (one from the male parent and one from the female parent), exchanges of segments of parental chromosomes (recombination) while they are paired, and an extra division without DNA replication (Figure 4.1b). Following sexual conjugation, haploid male and female gametes unite to form a diploid zygote, which then undergoes mitotic divisions to generate a new organism.

Figure 4.2 illustrates this in a form designed to bring out the eternal cycling of the germ-line cells. The rest of the body (soma) is but a temporary site of residence. The soma (that's you an me) supports the germ-line, but does not contribute to its information content. The dashed arrows in the figure indicate Darwin's idea, now largely discredited, that the soma contributes information to the germ-line. This will be further considered in chapter 20.

The process of cell division, be it mitotic or meiotic, is highly regulated. There are critical "check-points," where appropriate adjustments can be made if monitoring reveals unsatisfactory progress (Murray 1994; Page and Orr-Weaver 1996), and where appropriate hormones might transmit environmental cues. Gonadal degeneration would appear to be a most extreme example of an adjustment to a disruption of meiosis (Guyer 1902; Sutton 1903).

Darwin (1859 ch.8, 262) noted: "We have seen that the sterility of hybrids [produced by crossing members of two independent species], which have their reproductive organs in *an imperfect condition*, is a very different case from the [initial] difficulty of uniting [members of] two pure species, which have their reproductive organs perfect." For example, a horse and an ass, members of two allied but distinct species, have perfect gonads producing gametes which can unite to generate a hybrid mule. The latter appears vigorous, save in one aspect, the gonads are

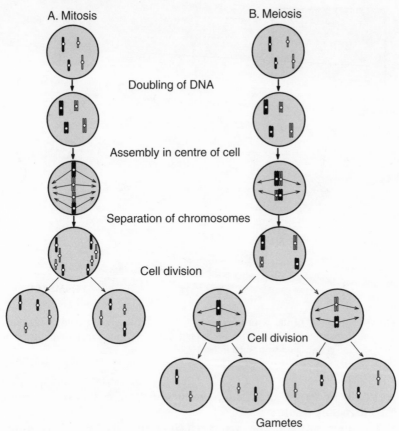

Figure 4.1
Outline of mitosis (A) and meiosis (B). Maternal chromosomes are white and paternal chromosomes are black. Recombination between homologous chromosomes is not shown.

poorly developed and gamete production is impaired. The mule suffers from gonadal degeneration and is sterile. In Huxley's words (1863 ch.11, 442) these: "hybrids ... are physiologically imperfect and deficient in the structural parts of the reproductive elements necessary for generation."

CATEGORIES OF REPRODUCTIVE ISOLATION

Turning again to Figure 4.2, it can be seen that the eternal cycling of the germ line can, in theory, be interrupted either directly at some stage of the cycle, or indirectly by removing the support provided by the soma. When you die, the germ cells within you die also. Unless your gametes are

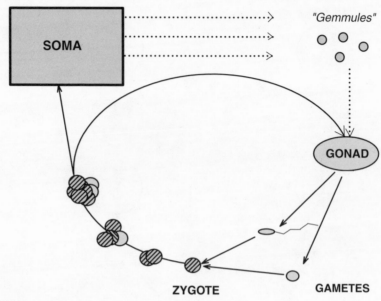

Figure 4.2
The eternal cycling of the germ-line. In male animals the gonad would be the testis
and the female gonad would be the ovary. Darwin's (incorrect) hypothesis of
"Pangenesis" is indicated by dashed lines.

artificially preserved, successful reproduction requires that you transmit
gametes before death.

The interruption of the cycling which occurs when members of different species do not successfully breed together, means that the species
are reproductively isolated from each other. Cycling is less likely to be
interrupted when members of one species attempt to interbreed. Hybrid
sterility and the accompanying gonadal degeneration is the barrier to fertility most likely to prevent cycling in the case of closely allied species.
Other barriers are likely to have intervened in the case of less closely
related species (i.e., species which were once "allied," but have since
diverged further from each other). Three general causes of reproductive
isolation were recognized by Darwin:

(i) *Transfer barrier.* For a variety of reasons (e.g., geographic separation, psychological incompatibility, anatomical difference, gamete
incompatibility) the gametes may not be able to meet or, having met, to
fuse to form a zygote ("pre-zygotic" reproductive isolation). Darwin
(1859 ch.4, 103) considered this might derive: "From haunting different
stations, from breeding at slightly different seasons, or from the individuals of each variety preferring to pair together."

Using plants as an example, Darwin (1859 ch.8, 263) pointed out: "There must sometimes be a physical impossibility in the male element [pollen] reaching the ovule [female element], as would be the case with a plant having a pistil too long for the pollen tubes to reach the ovarium."

(ii) *Developmental barrier.* The zygote may be unable to complete the cell divisions, or the cell differentiations (to generate tissues), or the organizational steps (to generate organs), which are needed for development of the embryo and its growth into the adult form. Thus Darwin noted (1859 ch.8, 264): "An embryo may be developed, and then perish at an early period … . I believe from observations … in hybridizing gallinaceous birds, that the early death of the embryo is a very frequent cause of sterility in *first crosses.*" In this case, the soma is not available to support the activities of the germ-line cells, even if the latter are not themselves affected by whatever has impaired the development of the soma. This "*hybrid inviability*" is one type of "post-zygotic" reproductive isolation.

(iii) *Gonadal barrier.* Hybrid sterility is another type of post-zygotic isolation, which should be distinguished clearly from hybrid inviability (Darwin 1859 ch.8, 245-6): "In treating this subject, two classes of facts, to a large extent fundamentally different, have generally been confounded together [confused]; namely, the sterility of [members of] two species when *first* crossed [developmental barrier; hybrid inviability], and the sterility of the hybrids produced from them [gonadal barrier; hybrid sterility]."

In his *Notes on Hybridity* (1862, 707) Darwin recognized that one barrier might occur first and then be replaced by another: "I suspect … [that hybrid] sterility must be habitually acquired *before* instinctive dislike [pre-zygotic isolation] … ; otherwise this instinctive dislike between potentially pairing members of opposite sexes would have been potent by itself and sterility [post-zygotic isolation] would have been superfluous."

Thus, if we return to our example of the prototypic horse and giraffe (chapter 2), we see the modern species with an obvious transfer barrier (inability to copulate). Pre-existing barriers (developmental, gonadal) might still exist, but selection pressures for their retention (relating to their function as reproductive barriers) are absent. As we go back in time, towards the divergence point (Figure 2.4), we arrive at two forms where copulation and formation of a zygote are possible. However, it is likely that the zygote is not able to overcome the developmental barrier, so that no adult organisms are produced (hybrid inviability). As we go further back in time, even closer to the divergence point, we find the developmental barrier is overcome, so that prototypic adult forms are produced. Here we see only the one and most fundamental barrier, the gonadal barrier (hybrid sterility) between these "allied species." (For reasons to be discussed in Chapter 6, the gonadal barrier is referred to in

Figure 2.4 as the chromosome pairing barrier.) Finally, we arrive at the actual divergence point itself. Here there is just one population whose members freely interbreed.

SUMMARY

Sterility comes in all shapes and sizes. It is not surprising that the Victorians were confused. Distinctions are made here between the various sterilities, with a focus on hybrid sterility and the associated gonadal degeneration. This requires consideration of the eternal cycling of the germ-line ("continuity of the germ-plasm"), meiosis, and the various categories of reproductive isolation.

5 Physiological Selection

"When you have eliminated the impossible, whatever remains,
however improbable, must be the truth."
Sherlock Holmes in *The Sign of Four*
by A. Conan Doyle (1890)

Whether or not George Romanes was influenced by Conan Doyle, certainly he was influenced by Charles Darwin, who became his close mentor. The book *Isolation and Physiological Selection* (1897), the last of Romanes' three volume series *Darwin, and After Darwin*, must surely rank, next to Darwin's *The Origin of Species by Means of Natural Selection* (1859), as one of the great scientific detective stories of all time. Extending speculations of Belt (1874) and Catchpool (1884), the Romanes–Gulick theory of the origin of species by means of "physiological selection" went as far as the Victorians could be expected to have gone in identifying the culprit, given the state of biological knowledge at the time. In Part 2 of the present book, I reveal the chemical identity of the "physiological peculiarity" which may be responsible for the most fundamental form of reproductive selection, as manifest in the phenomenon of hybrid sterility.

As set out in chapter 3, some of Darwin's most eminent critics had expressed the view that natural selection was not enough. Conventional natural selection (phenotypic selection) could transform a single species so that "monotypic" evolution would occur, but without some form of reproductive selection (isolation) to prevent intercrossing, branching into two species ("polytypic evolution") could not occur (Figure 5.1). Romanes (1886) wrote that:

Physiological selection, by preventing such intercrossing, *enables* natural selection to promote diversity of character, and thus to evolve species in ramifying branches instead of in linear series. ... Cross-sterility in species cannot possibly be due to natural selection; but everywhere arises as a result

of some physiological change having exclusive reference to the sexual system, a change which is probably everywhere due to the *same* cause ... .

The critical difference between the Romanes–Gulick and Darwin–
Wallace viewpoints is shown in Figure 5.2. Darwin held that phenotypic
("natural") selection was the prime moving force of evolution with
reproductive selection following "incidentally." This view continued to be
championed by Wallace after Darwin's death. On the other hand,
Romanes held that reproductive selection would occur *first*, or would be
a close concomitant of phenotypic selection. In the absence of one of the
other forms of reproductive selection (the most usual situation), physiological selection would be *the* critical isolating factor. In Romanes' words
(1897 ch.3, 46-7): "The older, and hitherto current theory, supposes the
cross-infertility to be but an *accident* of specific divergence, which, therefore, has nothing to do with *causing* the divergence. The newer theory, on
the other hand, supposes the cross-infertility to have often been a *necessary condition* to the divergence having begun at all" [all italics are
Romanes'].

A. Monotypic   B. Polytypic

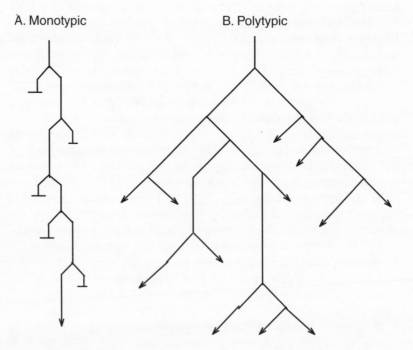

Figure 5.1
Distinction between serial, non-branching evolution, and divergent evolution.

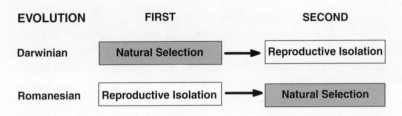

Figure 5.2
Distinction between Darwinian and Romanesian theories.

However, with relatively minor adaptations, "the older and hitherto current theory" of Darwin and Wallace still remains attractive to many evolutionists.

INTERNAL CONTRIBUTION

There had long been a feeling of the need for an *"internal"* contribution to the evolutionary process (see chapter 3). There was also a wonder at the discreteness of species, and the lack of an obvious adaptive value (the inutility) of many of the characters which distinguished species. In *Materials for the Study of Variation* (1894 Introduction, 11) Bateson observed:

> As Darwin and many others have often pointed out, the characters which visibly differentiate species are not as a rule capital facts in the constitution of vital organs, but more often they are just those features which seem to us useless and trivial, such as the patterns of scales, the details of sculpture on chitin or shells, differences in numbers of hairs or spines .... These differences are often complex and are strikingly constant, but *their utility is in every case problematical.*

He considered (Introduction, 17) that: "The first question which the study of variation may be expected to answer, relates to that discontinuity of which species is [are] the objective expression. Such discontinuity is not in the environment; may it not, then, be *in the living thing itself?*"

By "discontinuity" he meant the fact that variations characteristic of a particular species tend to be distinct; thus each species tends to be discrete with no intermediate forms between it and allied species, which have arisen with it from a common ancestral species. Bateson later (1909a, 230) affirmed Hooker's maxim that: "the 'Origin of Variation,' whatever it is, is the only true 'Origin of Species,'" and concluded (1913 ch.1, 30) that: "As soon as it was realized how largely the phenomena of variation and stability must be an index of the *internal constitution* of organisms, and not mere consequences of their relations to the outer world, such phenomena acquired a new and more profound significance."

As noted earlier, Lyell (1863 ch.21, 420) and others had pointed out that organisms may differ in their internal chemistry, but show no obvious morphological difference. As examples he pointed to the facts that: "In one [geographical] region a [plant] species may possess peculiar medicinal qualities which it [the same species] wants [lacks] in another region, or it may be hardier and better able to resist cold." Romanes (Figure 5.3) suspected that among these internal modifications in physiological constitution there might be a "peculiarity" in the reproductive system conferring reproductive isolation.

Figure 5.3
George Romanes (1848-94), circa 1890.

NON-ADAPTIVE CHARACTERS

Anatomical or physiological variations were held to occur randomly and to be either "adaptive" (positively or negatively) and thus a potential target for phenotypic (natural) selection, or "non-adaptive" and thus not a potential target for phenotypic selection. Darwin (1859 ch.4, 85) had noted: "the many small points of difference between species, which, as far as our ignorance permits us to judge, seem to be quite unimportant [for survival]."

He had explained these small differences in terms of: "many unknown laws of *correlation of growth*, which, when one part of the organisation is modified through variation, and the modifications are accumulated by *natural selection* ... will cause other modifications, often of the most unexpected nature."

Thus, Darwin acknowledged that some characters almost certainly conferred no adaptive advantage, and suggested this to be due to a "correlation of growth" with adaptive characters, adding that "the nature of the bond of correlation is very frequently quite obscure" (1859 ch. 5, 144). Darwin was not what we would now call a "neutralist" (see chapter 11).

Extensive independent studies on incipient species (varieties) and species by Nägeli, Wagner, Jordan, and Le Conte found that certain groups of plants showed only minute anatomical differences, yet were partially (in the case of "varieties") or completely (in the case of "species") reproductively isolated from each other. Romanes (1897 ch.5, 87-8) summarized these researches:

(1) The research embraces large numbers of species belonging to very numerous and very varied orders of plants; (2) in the majority of cases ... indigenous species which are of common occurrence present constant varieties; (3) these varieties, nevertheless, may be morphologically so slight [in differences from each other] as to be almost imperceptible [as varieties]; (4) they occupy common areas and grow in intimate association; (5) although many of them have undergone so small an amount of morphological change, they have undergone a surprising amount of physiological change; for (6) not only do very many of these varieties come true to seed [produce their like when crossed with the same variety]; but, (7) when they do, they are always more or less cross-infertile *inter se* [between different varieties].

That so many independent, yet closely allied (morphologically) species, could comingle in a common environment, prompted consideration of possible mechanisms of non-adaptive evolution (i.e., evolution *without* phenotypic selection).

DISCRIMINATE ISOLATION

Darwin had noted (1859 ch.4, 103) that: "Intercrossing plays a very important part in nature in keeping the individuals of the same species, or of the same variety, *true and uniform* in character." However, he did not fully explore the implications of the converse, namely the effect on phenotypic characters of an absence of intercrossing (reproductive isolation). Romanes (1897 ch.1, 2) held:

That in the principle of [reproductive] *Isolation* we have a principle so fundamental and so universal, that even the great principle of *Natural Selection* [phenotypic selection] lies less deep, and pervades a region of smaller extent. Equalled only in its importance by the two basal principles of *Heredity* and *Variation*, this principle of *Isolation* constitutes the third pillar of a tripod on which is reared the whole superstructure of organic evolution. ... By isolation I mean simply the prevention of inter-crossing between [members of] a separated section of a species or kind and [members of] the rest of that species or kind.

Romanes (ch.1, 5) made a distinction between discriminate and indiscriminate isolation:

If it be discriminate, the isolation has reference to the resemblance of the separated individuals to one another; if it be indiscriminate, it has no such reference. For example, if a shepherd divides a flock of sheep without regard to their characters, he is isolating one section from the other *indiscriminately*; but if he places all the white sheep in one field, and all the black sheep in another field, he is isolating one section from the other *discriminately*.

He then presented (ch.1, 8) the case for a major role of discriminate isolation in organic evolution:

To state the case in the most general terms, we may say that if the other two basal principles are given to heredity and variability, the whole theory of organic evolution becomes ... a theory of the causes which lead to discriminate isolation, or the breeding of like with like to the exclusion of unlike. For the more we believe in heredity and variability as basal principles ... the stronger must become our persuasion that discriminate breeding leads to divergence of type, while indiscriminate breeding leads to uniformity. ... So long as there is free intercrossing, *heredity cancels variability*, and makes in favour of [promotes] fixity of type. Only when assisted by some form of discriminate isolation, which determines the exclusive breeding of like with like, can heredity make in favour of [promote] change of type, or lead to what we understand by organic evolution.

This discriminate isolation would be achieved by means of reproductive selection (ch.1, 8-9):

Sexual incompatibility must be held to have taken an immensely important part in the differentiation of varieties into species. ... *Wherever* such incompatibility is concerned, it is to be regarded as an isolating agency of the *very first* importance. And as it is of a character purely *physiological*, I have assigned to it the name *Physiological Isolation*; while for the particular case where this general principle is concerned in the origination of specific types [species], I have reserved the name *Physiological Selection*.

He is here proposing to use the term "isolation" for reproductive selection in general, and "selection" for the type of reproductive selection of most importance for speciation events. In this book I generally use "isolation" and "selection" interchangeably (see chapter 2), but from the context it should be clear whether I am referring to the general or the particular.

Natural selection itself can provide a form of discriminate isolation permitting species adaptation (Figures 2.1, 5.1a). Romanes noted (1897 ch.1, 9):

> The other most important form of discriminate isolation to which I have alluded is *Natural Selection* ... [which] is the exclusive breeding of those better adapted to the environment: ... it is a process in which the fittest are prevented from crossing with the less fitted, by the *exclusion* of the less fitted. Therefore it is, strictly and accurately, a mode of isolation, where the isolation has reference to *adaptation*, and is secured in the most effectual of possible ways, i.e., by the destruction of all individuals whose intercrossing would interfere with the isolation.

INDISCRIMINATE ISOLATION AND RANDOM DRIFT

Returning to the example of sheep Romanes (1897 ch.1, 11) asked:

> Is it not self-evident that the farmer who separated his stock into two or more parts indiscriminately, would not effect any more change in his stock than if he had left them all to breed together? Well, although at first sight this seems self-evident, it is in fact untrue. For, *unless the individuals which are indiscriminately isolated happen to be a very large number,* sooner or later their progeny will come to *differ* from that of the parental type, or unisolated portion of the previous stock. And, of course, as soon as this change of type begins, the isolation ceases to be indiscriminate. ... A specific type [species] may be regarded as the average mean of all its individual variations, any considerable departure from this average being, however, checked by intercrossing. ... Consequently, if from any cause a section of a species is prevented from intercrossing with the rest of the species, we might expect that new varieties should arise within that section, and that in time these varieties should pass into new species.

This process, which we would currently best equate with "random drift," had been hinted at by Hooker (1860) who noted: "Variation is

Figure 5.4
John Gulick (1832-1926).

effected by graduated changes; and the tendency to varieties, both in nature and under cultivation, when further varying, is rather to depart more and more widely from the original type, than to revert to it." Thus: "The abstract principle called variation is enough *with time* to beget any amount of change" (Hooker's italics; 1862a, 130).

This was more precisely formulated by the Belgian mathematician Delboeuf (1877). Once a barrier divides a group (be it artificial or natural, extrinsic to individuals or intrinsic to individuals), then, depending on the size of the initial group and its intrinsic variability, sooner or later the sections of the group will become differentiated from each other. This differentiation can be greatly accelerated by natural selection (phenotypic selection) acting *discriminately* (Figure 5.1b). Romanes (1897 ch.2, 33) noted:

> Natural selection is the forcing heat, acting simultaneously on each of the separate branches which has been induced to sprout *by other means*; and in thus rapidly advancing the growth of all the branches, it is still entitled to be regarded as the most important *single* cause of diversification in organic nature, although we must henceforth cease to regard it as in any instance the *originating* cause, or even so much as the *sustaining* cause [since if a barrier is removed, without substituting another barrier, the diversification would reverse].

The Reverend John Gulick (Figure 5.4) had been led to a similar position as Romanes through detailed studies of the distribution of snails among valleys in the Sandwich islands, pondering: "Have we found one of the 'centres of creation'?" (Gulick 1872a). The valleys each appeared to offer an identical habitat yet provided a degree of *geographical* isolation so that the slowly migrating snails in a particular valley would preferentially interbreed. In a paper read to the Linnean Society (1872b), Gulick proposed:

If a bird should carry a leaf bearing two individuals of some species and drop it a mile beyond the limits already reached by other members of the species, they might there ... multiply for some tens of years before the first scattering individuals from the slowly advancing wave of migration would reach them. By this time, ... with a pre-existing tendency to rapid variation, a certain variety of form and colour might have partially established itself among them. The arrival of a few individuals representing the old stock would, amongst the multitudes of the new variety, have no influence in bringing back the succeeding generations to the original form.

Gulick found that he could roughly estimate the degree of divergence between the occupants of any two valleys by their degree of geographical separation. The observed variations appeared without utility to the organisms concerned. Although difficult to prove absolutely, Darwinian phenotypic (natural) selection did not appear to have played a role either in generating or in sustaining the variations (Gulick 1872, 1887, 1905 ch.1, 3-6; Gulick 1932 ch.14, 387).

TRADE-OFFS BETWEEN FERTILITIES

As I noted in chapter 3, Darwin did not see how a loss of fertility could benefit a variant individual who would thus be barred from mating with other non-variant individuals. However, if the *loss* of fertility towards non-variants coincided with a *retention* of fertility towards similar variants, then the trade-off between the fertilities might be to the benefit of the individual (with respect to net reproductive success). This was lucidly explained by Romanes (1897 ch.3, 42):

Of all parts of those variable objects which we call organisms, the most variable is the reproductive system; and the variations may carry with them functional changes, which may be either to the direction of increased or of diminished fertility. Consequently, variations in the way of greater or less fertility frequently take place, both in animals and plants; and probably, if we had adequate means of observing this point, we should find there is no one variation more common. But of course where infertility arises ... it immediately becomes extinguished, seeing that the individuals which it affects are less able (if able at all) to propagate and to hand on the variation. If, however, the variant, while showing some degree of infertility with the parent form, *continues* to be fertile as before when mated with *similar* variants, under these circumstances there is no reason why such differential fertility should not be perpetuated. Stated in another form, this suggestion enables us to regard many, if not most, species as the records of variations in the reproductive systems of their ancestors.

On this subject, the mathematician Moulton is quoted by Romanes (1897 Appendix B, 159-60):

It is so easy to confuse the survival of an *individual* with the survival of a *peculiarity* of *type* [race, variety or species]. No one has ever said that an *individual* is assisted by the possession of selective fertility: that is a matter which cannot affect his [personal] chance of life. Nor has any one said that the possession of selective fertility in an *individual* will *of itself* increase the chance of his having progeny that will survive, and in turn become the progenitors of others that will survive. Taken by itself, the fact that an *individual* is capable of fertility with *some only* [not all] of the opposite sex lessens the chance of his having progeny.

Moulton then points to the evolutionary conflict between the needs of the individual and those of the group to which it belongs, which would require appropriate trade-offs:

Whether or not he [personally] is more or less favourably situated than his *confreres* for the battle of life must be decided by the *total sum* of his peculiarities [variations]; and the question whether or not ... selective fertility will be a hindrance must be decided by considerations depending on the other peculiarities *associated with it*.

But when we come to consider the survival or permanence of a *type* or *peculiarity* [group rather than the individual], the case is quite different. It then becomes not only a favourable circumstance, but, in my opinion, almost a *necessary condition*, that the peculiarity should be associated with selective fertility. ...

If cross-infertility is so intensely disadvantageous to the individuals presenting it, it cannot have been that which made these individuals and their progeny survive. It is therefore a burden which they have carried. But we find it is more or less present in *all* the closely allied types that occur on common areas: therefore it must be a necessary feature in the formation of such types; for it cannot be an accident that it is present in so many. In other words, it must be the *price* which the individual and his progeny pay for their formation into a type [variety or species].

BIOLOGICAL PURPOSE

As will be discussed more fully in chapter 14 in the context of the best interests of "selfish genes," the *biological* purpose of individual survival is to be able to produce healthy fertile offspring. So it would be expected that factors which increase either the personal survival of the individual (so that reproduction can occur), or the reproductive effectiveness of the individual (so that mating maximizes the number and fitness of progeny), will be of selective value. If there is a conflict between personal survival and reproductive effectiveness, then the various selective forces involved will have approached an appropriate compromise (with the biological purpose as the overriding goal).

There is nothing strange in this. Reproduction requires gametes. The average human male produces 200 million sperms each day, which Braun (1998) calculates as 1,500 sperms per heart beat! Hormonally-driven libidinous desires, the menstrual bleeding of non-pregnant females, and the non-copulatory discharge of semen by males, periodically remind us of our underlying biological purpose. Indeed, the path to mental health seems to require that one comes to terms with one's biology. Romanes (1892 ch.7, 264-5) stated:

The life which it is the object, so to speak, of natural selection to preserve, is primarily the life of the *species*. Natural selection preserves the life of the *individual* only in so far as this is conducive to that of the species. Wherever the life-interests of the individual clash with those of the species, that individual is sacrificed in favour of others who happen better to subserve the interests of the species. For example, in all organisms a greater or less amount of vigour is wasted, so far as individual interests are concerned, in the formation and nourishment of progeny. In the great majority of plants and animals an enormous amount of physiological energy is thus expended. Look at the roe or the milt of a herring, for instance, and see what a huge drain has been made upon the individual for the sake of the species. Again, all *unselfish* instincts have been developed for the sake of the species, and usually against the interests of the individual. ... And, in a lesser degree, the parental instincts, wherever they occur, are more or less detrimental to the interests of the individual, though correspondingly essential for the race. ... But of course in many, if not the majority of cases, anything that adds to the life-sustaining power of the single life thereby ministers also to the life-sustaining power of the type [species]; and thus we can understand why all mechanisms and instincts which minister to the single life have been developed, namely, because the life of the species is made up of the lives of all its constituent individuals. It is only where the interests of the one *clash with those of the other* that natural selection works against the individual. So long as the interests are coincident, it works in favour of both.

We will not pursue the meaning of "interests of the individual," save to define it rather vaguely for humans as "life, liberty and the pursuit of happiness," to the exclusion of the aspect of "happiness" which involves reproduction. The Huxley quote at the beginning of the Prologue seems to express one personal view of happiness as an engagement in the unending search for "truth." Natural selection has so "wired" our heads that for many in contemporary society "happiness" and successful reproduction are not greatly distinguished.

THE "REPRODUCTIVE SYSTEM"

By "reproductive system" or "reproductive elements" we can assume the Victorians usually meant the gonad, rather than the secondary sexual characters required for conjugation and the nourishment of offspring. The gonad assumed a more central role following Weismann's clarification of the "continuity of the germ-plasm" (Figure 4.2; Weismann 1893 ch.1, 198-224; Romanes 1893). Darwin had held (1859 ch.1, 8-9) that: "the most frequent cause of variability may be attributed to the male and female reproductive elements having been affected prior to the act of conception, ... [and] we owe variability to the same cause which produces sterility." Romanes noted (1897 Appendix C, 169):

> For Weismann's theory supposes that all changes of specific type [affecting speciation] must have their origin in variation of a continuous germ-plasm. But *the more the origin of species is referred directly to variations arising in the sexual elements* [germ-plasm], *the greater is the play given to the principles of physiological selection* [Romanes' italics]; while, on the other hand, the less standing is furnished to the theory that cross-infertility between allied species is due to 'external conditions of life' [as argued by Darwin].

Thus, physiological selection was seen as something influencing the germ-line, the bodily seat of which was localized to the germ cells of the gonads. Physiological selection was discerned as an isolating agency, a physiological process, working through the slow random occurrence of variations affecting something in the gonads. Physiological selection was manifest as decreased fertility with some members of the species to which an individual initially belonged, while fertility with other members of the species (initially, perhaps a small group) *was maintained*. Romanes could only hint at the mechanism (1897 ch.3, 43-4):

Physiological isolation [selection] depends on distinctive characters belonging to organisms themselves; and it would be opposed to the whole theory of descent with *progressive* modification to imagine that absolute sterility usually arises in a *single* generation between two sections of a perfectly fertile species. Therefore evolutionists must believe that in most, if not in all cases (could we trace the history, say of any two species, which have sprung from a single parent stock in a common area, and are now absolutely sterile with one another), we should find that this mutual sterility has been itself the product of *gradual evolution*. Starting from complete fertility within the limits of a single parent species, the infertility between derivative or divergent species, *at whatever stage in their evolution this began to occur* [Romanes' italics], must usually at first have been *well-nigh imperceptible*, and thenceforth have proceeded to increase stage by stage. ... Does it not therefore become ... in a high degree probable, that from the moment of its inception this isolating agency must have played the part of a segregating cause, *in a degree proportional* to that of its completeness as a physiological character?

Romanes here envisages a slow, incremental process, occurring in the gonad and leading first to partial, and eventually to complete, infertility with the parent species. When provoked by Wallace, he had noted that selective fertility would require the "suitable mating of 'physiological complements'" (Romanes 1897 ch.5, 94).

"WILL NOT DO"

Under the influence of Hooker (1862a, 129), who urged him to distinguish between variation "as the origin of species, and natural selection as the *fixer* of these [species]" [Hooker's italics]), and Huxley (who pressed him on hybrid sterility), Darwin (1862) in *Notes on Hybridity* had himself struggled with what Romanes was later to call physiological selection. The idea of variation affecting the reproductive system *itself* is evident from remarks on pollen (702-8): "Pollen can be modified merely to favour crossing; with equal readiness it could be modified to prevent crossing. ... If we once admit the [ease] of differentiation of pollen, *we can understand hybridity*. ... [Natural] selection only collects [? takes advantage of] and forms [? builds on] this sterility" [Darwin's italics]. That variation in the reproductive system should accompany other variations was recognized (706):

If [a] new var[iant organism] paired with [a member of the] old [parental group], it would be lost: if paired with [a] similar var[iant] it would be preserved. Sterility must supervene at [the] first formation of [the] var[iant]. When we see how almost universal some sterility is, can we believe [it is] *always* acquired? [It] must be incidental [random]. When [the] whole body of a species changes [see Fig. 2], ... [there is] no use in [no need for] sterility [Darwin's italics].

However, he was unable to untangle the issues (708):

An animal, say, becoming adapted for aquatic life and surrounded by so many of its parent terrestrial forms, ... [because of crossing] would be dragged back from its favourable variation. But if a variation arose, which was sterile with the ordinary form [parental group], then although even so large a proportion united with ordinary form and were ultimately lost by sterility ... , if even so few united with similar form, these would be kept pure; but then *chances* are ... that they would [then] unite with [the] ordinary form and so would be lost by sterility. It will not do. ... If sexual disinclination [pre-zygotic isolation] supervened all would go well; but then why does sterility supervene, and why in [the] hybrid [second cross], and not always in [the] first cross? **Will not do** [Darwin's italics and bold emphasis].

Later, claiming agreement with Weismann (1875), he came closer to a Romanesian position (Darwin 1875 ch.23, 282):

When we reflect on these facts we become deeply impressed with the conviction that in such cases the nature of the variation depends but little on the conditions to which the plant has been exposed, and not in any special manner on its individual character, but much more on the inherited nature or constitution of the whole group of allied beings to which the plant in question belongs. We are thus driven to conclude that in most cases the conditions of life [environmental factors] play a subordinate part in causing any particular modification; like that which a spark plays, when a mass of combustibles bursts into flame – the nature of the flame [modification] depending on the combustible matter [Romanesian interpretation: reproductive isolation], and not on the spark [Romanesian interpretation: variation with or without natural selection].

Environmental factors would include natural selection. Variation when considered "spontaneous" would be independent of the environment, so that, without natural selection the "flame" (modification) would itself appear "spontaneously," independent of any environmental input (i.e., spontaneous phenotypic "combustion"). However, Darwin never went as far as Hooker (see chapter 3) in entirely dismissing environmental causes of variation. For example, Darwin (1878 ch.12, 447-8) noted:

The advantages of cross-fertilization do not follow from some mysterious virtue in the mere union of two distinct individuals, but from such individuals having been subjected during previous generations to different conditions, or to their having varied in a manner commonly called spontaneous, so that in either case their sexual elements [gonads] have been in some degree differentiated. ... The injury from self-fertilization [adverse effects of inbreeding] follows from the want of such differentiation in the sexual elements.

THE "MORE ULTIMATE PROBLEM"

Romanes (1886a) regarded his studies as merely paving the way for future investigations:

"My suggested explanation of the origin of species opens up another and more ultimate problem, namely, granted that species have originated in the way supposed, what have been the *causes* of the particular kind of variation in the reproductive system which the theory requires?" These words were echoed in his final work (1897 ch.4, 62-3):

First, let it be carefully observed that here we have to do only with the *fact* of selective fertility, and with its *consequences* as supposed by the theory: we have nothing to do either with its *causes* or its *degrees*. Not with its causes, because in this respect the theory of physiological selection is in just the same position as that of [the theory of] natural selection: it is enough for both if the needful variations are provided, without it being incumbent on either to explain the causes which produce them [all italics are Romanes'].

To give his Victorian audience a concrete example of *one* form of physiological isolation, Romanes (1886a) had mentioned a change in the time of flowering of a plant, which would restrict the plant to fertilization only by other members of the species which had undergone the same change in flowering time:

Suppose the variation in the reproductive system is such that the season of flowering or of pairing becomes either advanced or retarded. ... some individuals living on the same geographical area as the rest of their species, have varied in their reproductive system, so that they can only propagate with each other. They are thus perfectly fertile *inter se,* while absolutely sterile with all the other members of their species. This particular variation being communicated by inheritance to their progeny, there would soon arise on the same area, ... two varieties of the same species, each perfectly fertile within its own limits, while absolutely sterile with one another. That is to say, there has arisen between these two varieties a barrier of intercrossing which is quite as effectual as a thousand miles of ocean; the only difference is that the barrier, instead of being geographical, is physiological.

As discussed in the next chapter, we would now regard such cases as resulting from primary variations in specific genes which affect the time of flowering ("genic sterility"), thus creating a gamete transfer barrier. However, Romanes (1886a) sensed that something more than this was involved in the *general* case of formation of new species by physiological selection:

In many cases, no doubt, this particular ... variation, has been caused by the season of flowering or of pairing having been either advanced or retarded in a section of a species, or to sundry other influences of an *extrinsic* kind; but probably in a *still greater number of cases* it has been due to what I have called *intrinsic* causes, or to the "spontaneous" variability of the reproductive system itself.

The change in time of flowering was not associated with degeneration of the gonad, unlike hybrid sterility ("intrinsic" cause). Thus, pollen from an early-flowering plant, if saved, should successfully fertilize a late-flowering plant. Similarly, variants in pairing time (cicadas, moths) can be successfully crossed (Bateson 1913 ch.6, 119). Indeed, Bateson dismissed sterility due to certain anther variants in plants as being due to a "simple recessive character" (ch.11, 239), implying that the phenomenon (genic sterility) was unlikely to be of *general* significance for the evolution of species (see next chapter).

To put it mildly, Romanes' Physiological Selection Theory was not well received by his contemporaries (see Part 4). It did, however, receive prominent attention in Kellogg's *Darwinism Today* (1907). The latter book, while presenting a masterly summary of nineteenth century thought, was probably composed before the full importance of Mendel's observations was realized; to this extent, the book may not at that time have been considered what we would these days call "cutting edge" science.

SUMMARY

Having eliminated the impossible, Darwin's close research associate, George Romanes, explored the improbable. In the context of Delboeufian "random drift," he distinguished adaptive and non-adaptive characters, and the associated discriminate and indiscriminate forms of isolation. Spontaneous "intrinsic" variations within the gonads of some members of a species (physiological as opposed to morphological variations) decrease the fertility of crosses with members of the parental group, but retain fertility of crosses with members who have undergone the *same* variation. Such "physiological complements" (potential incipient species) are organisms with *compatible* reproductive systems, so that offspring can be produced. This "physiological selection" progressively increases the reproductive isolation of the incipient group, thus creating conditions favourable to the preservation of non-adaptive or adaptive phenotypic variations. The latter would constitute a target for natural selection. The spontaneous "intrinsic" gonadal variations provide a general basis for the origin of species, and are distinct from "extrinsic" variations affecting phenotypic characteristics such as time of pairing.

6 Failure of Meiotic Pairing

"Before the problems of heredity and development Bateson always maintained sublime steadiness of judgement. ... He asked questions which at that time were unanswerable and had the courage to develop the consequences of his convictions."

William Coleman 1970

Although given a variety of names (physiological units, pangens, biophors, character units), the idea that inherited phenotypic characteristics were determined in some way by distinct elements, now equated with "genes," was well established by the end of the nineteenth century (Kellogg 1907). With the re-emergence of the work of Mendel (1865) and new microscopical observations of cells, the time was ripe for theoretical synthesis. Because it explained so much, it was easy (and certainly politically correct; see Part 4) to believe that the "genic" viewpoint explained everything. Thus, theories tended to be couched entirely in terms of genes. Bateson was one of the few who thought otherwise.

"INGREDIENTS" OR "FACTORS"

Microscopic studies of the degenerate gonads of sterile hybrids (Guyer 1902; Montgomery 1902; Sutton 1903; Federley 1913) were summarized by Bateson (1917, 205; Figure 6.1):

> The sterility common among hybrids is in part a consequence of their inability to *sort out* into gametes the *ingredients* which are united in them. Such hybrid zygotes cannot then make gametes, and consistent with this view it is often found in these cases that the germinal processes have a normal course until the reduction division [of meiosis] should occur, when the nuclear materials fail to divide properly and deformed cells result.

Bateson felt that the "ingredients" were more than just particles, sensing that their structure should in some way reflect what we would now call

Figure 6.1
William Bateson (1861-1926).

their "information" content. Perhaps he had read his Aristotle. The latter observed that the "eidos," the form-giving essence that shapes the embryo, "contributes nothing to the material body of the embryo but only communicates its program of development." Bateson noted (1908, 321) that:

> Several of them [ingredients or factors] behave much as if they were ferments [enzymes catalyzing chemical reactions], and others [behave much] as if they constructed the substances on which the ferments act [i.e., they were substrates]. But we must not suppose for a moment that it is the ferment, or the objective substance [substrate], which is transmitted [from generation to generation]. The thing transmitted can only be the power or faculty ["information" to the modern reader] to produce the ferment or objective substance.

Shortly thereafter, the words "genotype" and "phenotype" were coined (Johannsen 1911). At the Melbourne meeting of the British Association Bateson noted (1914, 280):

> The allotment of characteristics among offspring is ... accomplished ... by a process of *cell division*, in which numbers of these characters, or rather the elements on which they depend, are sorted out among the resulting germ cells in an orderly fashion. What these elements, or *factors* as we call them, are, we do not know. That they are in some way directly transmitted by the material of the ovum and of the spermatozoon is obvious, but it seems to be unlikely that they are in any simple or literal sense material particles. I suspect rather that their properties depend on some phenomenon of arrangement [all italics are Bateson's].

Here he may have had an intuition of the coding properties of a sequence, since when considering how the factors might vary to cause changes in observable characters, he further noted: "That which is conferred in variation must rather itself be a change, not of material, but of arrangement, or of motion. ... By the re-arrangement of a very moderate number of things we soon reach a number of possibilities practically infinite."

J.B.S. Haldane later (1957) affirmed that Bateson: "never accepted the word 'gene' with its rather wide connotations. ... Bateson used the neutral word 'factor' ... [which] can be anything from a difference of a few atoms in a single nucleotide, to an inversion or the presence of an extra chromosome; for these too are inherited in a Mendelian manner. If this is so, ... all evolution is the accumulation or loss of [such Batesonian] factors." Bateson (1907) was also wary about jumping to the obvious conclusion that all hereditary characters were borne by chromosomes:

> All that has been observed by cytologists is consistent with the results of experimental genetics. The recognition of a definite differentiation among the chromosomes (see especially Sutton [1902]) is probably an important advance, though until we can positively recognize characters in the zygote as associated with some visible cytological elements we must be cautious in forming positive conclusions ... . There is still no positive proof that segregation [of Mendelian characters] occurs at the reduction division [of meiosis], but all the facts known point to that conclusion. Indeed we can scarcely doubt that this is the critical moment. Investigations of sterile forms (Guyer [Dissertation 1900]; Gregory [1905]) show that it is often at this division that abnormality begins.

COMPLEMENTARY FACTORS INHIBIT MEIOSIS

With the above understanding of the term "factor" in mind, we will next consider an important statement of Bateson (1909a, 228-9), which we will analyze carefully in chapter 13:

> When [members of] two species, both perfectly fertile [with members of their own species] ... , produce on crossing a sterile progeny, there is a presumption that the sterility is due to the development in the hybrid of some *substance* which can only be formed by the meeting of *two complementary factors* ... . Now if the sterility of the cross-bred [hybrid] be really the consequence of the meeting of *two* complementary factors, we see that the phenomenon could only be produced among the divergent offspring [of a cross between members] of *one* species [Bateson's italics] by the acquisition of at least *two* new factors; for if the acquisition of a *single* factor [by an individual] caused sterility *the line would then end*. Moreover

each factor must be *separately* acquired by *distinct individuals* [prior to a cross between them], for if both were present *together*, the possessors would by hypothesis be sterile.

And in order to imitate the case of species, each of these factors must be acquired by [members of] *distinct breeds*. The factors *need not*, and probably would not, *produce any other perceptible effects* ... . Not till the cross was actually made between the two complementary individuals would either factor come into play, and the effects even then might be *unobserved* until an attempt was made to breed from the cross-bred.

Bateson is here postulating factors arising in different individuals of a "species" which complement to produce a *negative* effect, hybrid sterility, but otherwise do *not* produce an obvious phenotype. Thus they would *not* seem to correspond with conventional characters which might (i) serve as targets for conventional Darwinian selection, (ii) be used in breeding studies (Mendelian analysis), and (iii) thus be shown to map in linear order on chromosomes. Bateson and co-workers showed that some of these conventional genetic characters tend to be inherited together (coupled in gametes in what we now refer to as a linkage group; Bateson 1907; Stern, 1950), and the technology for their mapping on individual chromosomes was developed by Morgan's group; (Sturtevant 1913). Continuing his remarks, Bateson pointed to the need for experimental studies of fertility (1909a, 229):

If the factors responsible for [hybrid] sterility were acquired, they would in all probability be peculiar to certain individuals and would not readily be distributed to the whole breed [entire species]. Any member of the breed also into which *both* factors [Bateson's italics] were introduced would drop out of the pedigree by virtue of its sterility. Hence the evidence that [members of] the various domesticated breeds ... can when mated together produce fertile offspring, is beside the mark [irrelevant]. The real question is, Do they *ever* produce sterile offspring? [i.e., Do separate complementary factors ever arise in two individual members of a species of opposite sex?] I think the evidence is clearly that sometimes they do [produce sterile offspring], oftener perhaps than is commonly supposed.

Here, indeed, would be "exceptions" which Bateson would "treasure." In seeking to spell out an agenda for further research, he later (1913 ch.11, 242-7) reiterated these points:

Though we cannot strictly define species, they yet have properties which varieties have not, and ... the distinction is not merely a matter of degree. The first step is to discover the nature of the factors which by their *complementary action* inhibit the critical [meiotic] divisions and so cause the sterility of the hybrid. ... But while following one plan or the other we shall still be awaiting the answer ... to the question whether among the various types [members of a species] there are some [exceptions] which differ from the rest in a *peculiar* way: whether by having *groups of characters* linked together in especially durable combinations, or by possessing ingredients which cause greater or lesser disturbances in the process of cell division, and especially the process of gametic maturation, when they are united by fertilization with complementary ingredients.

THE "RESIDUE": A SECOND LEVEL OF GENETIC INFORMATION

Still struggling with the problem, in a 1922 Toronto address to the American Association for the Advancement of Science Bateson declared, in the context of domesticated races, that he was:

... left with the conviction that some part of the chain of reasoning is missing. ... [Genetic] analysis has revealed hosts of transferable [classical Mendelian] characters. Their combinations suffice to supply in abundance series of [pheno]types which might pass for new species, and certainly would be so classed if they were met with in nature. Yet critically tested [by hybridizing], we find that they are not distinct species and we have no reason to suppose that any accumulation of characters of the same order [transferable characters amenable to conventional genetic analysis] would culminate in the production of distinct species.

So far he seemed to be covering old ground, but then he continued with a remark which must have appeared bizarre to those listening, but was merely repeating a point he had been reiterating for two decades: "Specific difference [that which is responsible for speciation] therefore must be regarded as probably *attached to the base* upon which these transferables *are implanted*, of which we know absolutely nothing at all."

What does this mean? It is clear from the preceding statement that "transferables" are the phenotypic characters which the breeder (or nature) selects, and thus are transferred to the offspring. The "base" is something to which things can be "attached" or "implanted," just as a statue has a base upon which it rests. *Both* conventional phenotypic characters, *and* that which is responsible for species differences, can be assembled on this base. Thus the "base" can carry two kinds of information, that for phenotypic characters (let us say "primary" information) and that for species characters (let us say "secondary" information). Of course, it is not possible that in 1922 Bateson meant "base" in the sense of a base in DNA.

"Molecules" for the Victorians meant the small molecules which had recently been described, such as urea. They appreciated that such molecules might form larger aggregates, but they tended not to use the term "molecule" or "macromolecule" to describe these aggregates. Thus, it is likely that Bateson's conception of the chemical nature of hereditary information was quite close to that of the Dutch botanist, Hugo de Vries, with whom he had a warm professional relationship. In his *Intracellulare Pangenesis* (1889, 49) de Vries noted: "Therefore the material bearers of hereditary characters cannot be identical with the molecules of chemistry; they must be conceived of as units, built up from the latter, much larger than they, and yet invisibly small."

De Vries was one of those responsible for the rediscovery of Mendel's (1865) quantitative studies of inheritance demonstrating the existence of integral (unit) characters which did not blend in a hybrid. Greatly stimulated by this, Bateson and Saunders (1902) developed the view that there might be *two* forms of hereditary information, one *changeable* from generation to generation *within* a species and not directly involved in speciation, and one *unchangeable within* a species and directly involved in speciation:

Has a given organism a fixed number of unit-characters [genes to the modern reader]? Can we rightly conceive of the whole organism as composed of such unit characters, or is there some residue – a basis – upon which the unit characters are imposed? We know, of course, that we cannot isolate this residue from the unit characters. We cannot conceive of a pea, for example, that has no height, no colour, and so on; if all these were removed there would be no living organism left. But while we know these characters can be interchanged, we are bound to ask is there something not thus interchangeable? And if so, what is it? We are thus brought to face the further question of the bearing of the Mendelian facts on the nature of Species. The conception of Species, however we may formulate it, can hardly be supposed to attach to allelomorphic or analytical varieties [genes]. We may be driven to conceive "Species" as a phenomenon belonging to that "residue" spoken of above, but on the other hand we get a clearer conception of the nature of sterility in crossing.

The link with meiosis was then made explicit and included a footnote to the studies of Guyer cited above:

Though some degree of sterility in crossing is only one of the divers properties which may be associated with Specific difference, the relationship of such sterility to Mendelian phenomena must be a subject for most careful enquiry. ... We know of no Mendelian case [study of the segregation of conventional Mendelian characters] in which fertility is impaired [remark qualified in 1904; see below]. We may, perhaps, take this as an indication that the sterility of certain crosses is merely an indication that *they cannot divide up the characters among their gametes* [Bateson's italics]. If the parental characters, however dissimilar, can be split up, the gametes can be formed, ...

That the sterility of hybrids is generally connected in some way with inability to form germ cells correctly ... is fairly clear, and there is in some cases actual evidence that this deformity of the pollen grains of hybrids is due to irregularity or imperfection in the processes of division from which they result.

NON-MENDELIAN PHENOMENA OF HEREDITY?

Bateson later drew a distinction between sporadic cases of sterility associated with withered anthers in crosses between closely related plants, and cases of hybrid sterility regularly associated with plant or animal "mules" (Bateson, Saunders and Punnett 1904):

[Withered] anthers were seen from time to time in many families, though commonly confined to individual flowers. This sporadic sterility has not been particularly studied. It is of interest to compare this example of the definite appearance of sterility ... with the familiar occurrence of sterility in cross-breds [hybrids]. Such a phenomenon [hybrid sterility] has often been supposed to indicate *remoteness of kinship*, yet here a closely comparable effect occurs ... as the result of a cross between two types which must be *nearly related*. Mr. Gregory in a careful examination of the pollen-genesis [in the case of the nearly related types], found that the divisions were normal up to the reduction division, when the chromosomes formed shapeless knots and entanglements, failing to divide [i.e., a chromosomal morphology similar to that seen in hybrid sterility].

These points were elaborated by Bateson (1907):

Mendelian segregation proves the unity of [conventional] characters. Specific differences [the observed differences between species] we must suppose, are built up of characters. [However] Is it a sound deduction that specific differences come into existence [that species originate] by the addition or elimination of such character units? ... We must ask [first] whether ... certain kinds of differences segregate [as Mendelian unit characters] and that certain other kinds do not segregate; and secondly whether we shall then recognize that it is to the non-segregating that the conception of species [the clue to the origin of species] attaches with the greater propriety. ...

Of the *non*-Mendelian phenomena of heredity we know as yet almost nothing [Bateson's italics]. ... The phenomenon of sterility certainly counts for much in this part of genetics. As to the nature of this sterility and its limitations, even as to the rules of inheritance in those cases where sterility is partial, we have scarcely any adequate knowledge, yet such knowledge must be obtained as a necessary preliminary to a reliable judgement on the limitations of specific distinction. ... We may indeed ... be driven to conceive specific difference as a property of the residue or basis upon which the allelomorphic characters are implanted; but it is not easy to suppose that the [conventional] features, [unit characters such as] breadth of leaves, and length of flowering stem, ... are of this fundamental nature.

Although Bateson was not aware of it, he had been crossing ground which Romanes had covered two decades before. Much of the time Romanes and Bateson use the word "complementary" in converse contexts. Romanes' "physiological complements" are organisms with reproductive systems which are *compatible*, so that offspring will be produced (see chapter 5). Bateson's "complementary factors" are components of the reproductive systems which are *incompatible*, so that offspring will *not* be produced. However, Bateson later noted (1913 ch.11, 241) that organisms were incompatible either because the reproductive system of "each is lacking in one of two [Romanesian] complementary factors, or that each possesses a [Batesonian] factor with an inhibitory effect." Beyond this the Victorians and their Edwardian successors, who had no information on the chemical basis of variation and heredity, could not go. As we shall see (Part 2), "physiological complements" and "complementary factors" are but two sides of the same coin.

SUMMARY

Batesonian "factors" or "ingredients" contain hereditary information which may be encoded as a "phenomenon of arrangement." Within a species, some individuals develop factors with complementary properties. When two of these factors associate through sexual reproduction within a new individual, they cooperate to disrupt meiosis within the gonad. The resulting hybrid sterility in that individual is of importance for the origin of a new species. The complementary factors have no other phenotype, and are "attached to" or are a "property of" a "base" or "residue" upon which factors encoding conventionally segregating characters are also "implanted." The latter factors may change without changing the species. The complementary factors associated with the "base" tend to be unchangeable *within* a species, so that when they change, new species may result.

7 Conjugation of the Chromosomes

"The sterility of the offspring is in inverse proportion to the power of conjugation in the chromosomes."

ÖjvindWinge (1917)

Bateson's 1922 Toronto address generated a flurry of correspondence in both the general and the scientific literature. His reasoning was found "very difficult to understand, and his "idea of a specific base distinct from specific characters" was dismissed as "merely false metaphysics" (Cunningham 1922). One correspondent, however, came very close to grasping Bateson's message concerning the mechanism of chromosome pairing (Crowther 1922):

> If a sword and its scabbard are bent in different directions, it will happen sooner or later that the sword cannot be inserted, and the result will be the same whether the bending be effected by a single blow, or whether it be, in Dr. Bateson's words, "a product of a summation of variations." Is this illustration inapt? The sword and its scabbard are the homologous chromosomes. These presumably have to co-operate to produce the somatic cell of the hybrid, and their co-operation might be expected to require a certain resemblance, but for the production of sexual cells [gametes] they must do more, they must conjugate [pair]: and for conjugation it is surely reasonable to suppose that a much more intimate resemblance would be needed. We might, therefore, expect ... that, as species ... gradually diverged, it would be increasingly difficult to breed a hybrid between them: but that, even while a hybrid could still be produced, a fertile hybrid would be difficult or impossible, since the cells of the germ-track would fail to surmount the meiotic reduction stage when the homologous chromosomes conjugate.

Bateson (1922b) agreed noting: "It is not difficult to 'imagine' inter-specific sterility produced by a gradual (or sudden) modification. That sterility may quite reasonably be supposed to be due to the inability of certain chromosomes to conjugate, and Mr. Crowther's simile of the sword and the scabbard may serve to depict the sort of thing we might expect to happen."

Evidence that the failure of pairing was indeed likely to be "mechanical," in the sense of being due to a lack of chemical complementarity between chromosomes, was provided by studies of the phenomenon of polyploidy.

POLYPLOIDY "CURES" HYBRID STERILITY

Microscopic studies of chromosomes showed that the chromosome numbers of some closely related species were numerically related. Thus, three species of barley have 7, 14 and 21 pairs of chromosomes. This suggested that sometimes evolution would involve duplication and triplication of an original diploid chromosome set of 7 pairs.

Although a cross between members of two allied diploid species (producing haploid gametes) results in hybrid sterility, a cross between members of tetraploid versions of the two species (producing diploid gametes) is often quite *fertile*. The species barrier can be breached in these "allotetraploids" because pairing requires just *two* chromosomes, and many pairs can be accommodated during meiosis. In allotetraploids each chromosome at meiosis has a pairing partner (that which accompanied it in the gamete) and chromosomes can segregate normally to produce more diploid gametes (Muller 1914; Wright 1914; Winge 1917; Darlington 1932 ch.7, 182-96; Ohno 1970 ch.16, 101).

This amply demonstrates the importance of correct pairing for fertility (Figure 7.1), and provides a test system for sorting out whether failure of pairing is due to either a defect in some cofactor (a gene product) *required for the pairing* (which would tend to be defective both in diploids and tetraploids), or a defect in chromosomal pairing *itself* (a "structural" or "mechanical" incompatibility between chromosomes), which would be "cured" in the allotetraploid. Thus, chromosomes from the mother can pair with each other, and chromosomes from the father can pair with each other. If the sterility had been due to a functional defect in particular gene products, this "cure" would be unlikely (Dobzhansky 1937 ch.9, 290-1).

It should be noted that gene products are usually encoded by individual genes, so if a gene-product is functionally defective, the corresponding gene should have some structural defect (e.g., see the sample sequence in chapter 1 where one base change is sufficient to change the amino acid encoded). This change in primary information might *locally* affect, to a small degree, the pairing of the chromosomal region containing that gene with the corresponding region of a homologous chromosome containing

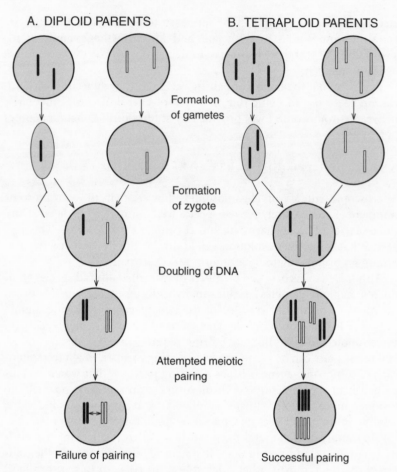

Figure 7.1
Formation of interspecies hybrids in the case of (A) diploids and (B) tetraploids.

the corresponding healthy gene (allele). This local impairment (due to local heterozygosity) should be insignificant with respect to the failure of *entire* chromosomes to pair (i.e., there has been no gross change in secondary information), and so should not itself be responsible for hybrid sterility.

STRUCTURAL INCOMPATIBILITY IN POLYPLOIDS

As described above, chromosomes in polyploid organisms may not pair for *functional* reasons, namely because of defects in gene products required to support the pairing process. Polyploid chromosomes can also fail to pair for *structural* reasons. Tetraploids can be produced *within* species ("autotetraploids"). In this case each chromosome has *three* homologous potential pairing partners, so that one part of a chromosome may pair with one partner, another part with a second partner and another part with a third partner. Thus, instead of "divalent" chromosome pairing (maternal with maternal, paternal with paternal), "tetravalent" pairing occurs. This usually disrupts meiosis.

Provided that a tetraploid mate (individual) can be found, an allotetraploid (individual) will breed true producing offspring with a blend of the phenotypic characteristics of both parent species. However, a cross of a tetraploid with a diploid results in a triploid organism with failure of some chromosome pairings. Tetraploids are *from the moment of their creation* reproductively isolated from the parental (diploid) species. This "instant" speciation process is one well understood mechanism for the production of new species. However, it is not the predominant mechanism. It is not *the* origin of species. One good reason for believing this is that diploids from two allied species must *pre-exist* for an allotetraploid to be formed. Since allied, the formation of the parental diploid species would have required a prior, and hence more fundamental, evolutionary divergence from a single ancestral species.

The cells of tetraploid organisms, and sometimes the organisms themselves (in the absence of a functioning cell number compensation mechanism), are larger than those of the corresponding diploids. Many novel hybrids produced by horticulturalists are viable polyploids, but the generation of polyploids in animals is difficult due to disparities between sex chromosomes (Muller 1925; Orr 1990; Forsdyke 1995b).

"GENIC" VERSUS "CHROMOSOMAL" INCOMPATIBILITY

Different scientists who independently make the same discovery often use different words to describe the same thing. This semantic difference can impede communication between the codiscoverers, and delay assimilation of the new knowledge by others. It was not until the 1920s that the key distinction between types of gonadal sterility made independently by Romanes and Bateson approached semantic clarification. Sterility could be due to a "genic" defect. If a sterility of this nature were suspected, scientists would seek to know more about the gene and the function of the product it encodes. Sterility could be "chromosomal," due to a defect in chromosome structure. If sterility of this nature were suspected, scientists would seek to know more about the nature of the structural defect.

Haldane (1927, 38-9) clearly distinguished the two potential causes of sterility, and suspected that chromosomal sterility would arise from differences in chromosome number.

The most serious argument against [natural] selection ... is that it does not explain the origin of interspecific sterility, except where it is due to external causes [pre-zygotic isolation] such as differences of size or breeding time. It is on these grounds that Bateson ... has criticized natural selection. But I have pointed out elsewhere [see chapter 19], a difference of a single gene between two animals may cause the production of an excess of one sex on crossing ... ; and several such genes may well cause complete sterility [sterility of polygenic origin]. Moreover, there is a second type of inheritable variation, leading to a change in chromosome number, which causes inter-varietal sterility, often without a very marked change in external characteristics.

At the time of Bateson's death, new information on the nature of mutations was emerging (Belling 1925; Muller 1928). This was summarized by biochemist Addison Gulick (1932 ch.16, 497-8):

Chromosomal incompatibilities are responsible for the formation of 'mules' of sterile hybrids, and in some cases for the failure of hybrid embryos at an early stage of development [hybrid inviability]. Since the chromosomes can coexist in the embryo and the adult hybrid, an incompatibility between them need not always interfere with the healthy life of the hybrid individual [no hybrid inviability]. But when the germ cells of the hybrid reach the stage of [meiotic] synapsis through which they must go before they can fertilize or be fertilized [become gametes], the chromosomes in them have to come into closer relations, and if incompatible elements are present in them they behave irregularly, generally agglutinate into a disorganized mass, and the cell [potential gamete] perishes before it is ready for fertilization.

He then proceeded to demonstrate (ch.16, 498) one mechanism by which chromosomal incompatibilities might result in reproductive isolation (Figure 7.2):

One of the newest findings in the elucidation of sterility between [organisms with] mutations has to do with interchromosomal translocations, in which two non-homologous chromosomes exchange part of their heredity-controlling units. Such a mutated stock, when it has become homozygous, is entirely fertile *with itself*; but when bred back to the parent-stock it may produce a sort of mule, half of whose germ cells are unable to develop [due to random assortment of chromosomes]. In the case of multiple translocations the viable germ cells of the mule will apparently be $(1/2)^n$ of the whole number, n being the number of chromosomal translocations. There is no obvious limitation for the size of n, and hence it is within the logical possibilities for virtually complete mule sterility to be established in this way. It is not known whether or not true mules are due to some such process.

Haldane's colleague Cyril Darlington noted in *Recent Advances in Cytology* (1932 ch.6, 156) that "Failure [of pairing] is not in itself evidence of dissimilarity of the chromosomes concerned, although pairing is evidence of similarity." To the rhetorical question, "What kinds of dissimilarity inhibit ... chromosome pairing?" Darlington replied (1932; chapter 6, 159):

There is evidence of two kinds of dissimilarity or differentiation in the chromatin material. First there is a qualitative differentiation between particles ... [with a] ... property of ... giving rise to differences (mutations) of a specific character, which has led to such an identifiable particle being described as a "gene". It is probably through gene mutations that the *qualitative* (intramolecular) differentiation arises. Secondly, there are [structural] differences due to *change in arrangement* or structure of these specific particles; these intermolecular changes determine genetic differences when they lead to changes in quantity, proportion and possible position.

Having no knowledge of DNA, he was using the terms "intramolecular" and "intermolecular" in a very loose sense. He went on (ch.6, 159) to argue that small dissimilarities were less likely to inhibit chromosome pairing than large dissimilarities:

Since intramolecular [genic] changes affect single molecules and intergenic [structural] changes affect larger or smaller groups of genes (perhaps many hundreds), though both [types of change] may inhibit pairing (on the simple assumption that only identical genes may pair), the first will have a negligible effect as compared with the second. Therefore the assumption will be made that all changes effective in reducing pairing, apart from changes in number, are intergenic or structural. Structural changes need not in themselves mean any change of genetical properties in the organism, except in a mechanical sense, since the same materials may be arranged in a different way without any phenotypic effect.

Dobzhansky (1937 ch.9, 261-87) reiterated the distinction between functional (genic) and structural defects and the way they affect chromosome pairing: "The essential fact is that a *failure* of pairing between chromosomes of different species is observed at meiosis in most ... sterile hybrids. ... The failure ... may, theoretically, be due to their structural dissimilarities, that is,

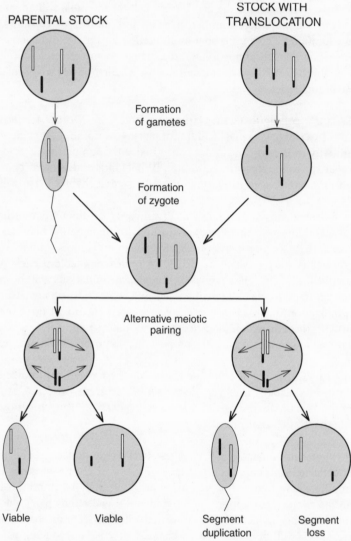

Figure 7.2
Addison Gulick's translocation model for speciation. In a cross between stock with a translocation and the parental stock, half the gametes are viable when there is one interchromosomal translocation (n=1). Only a quarter would be viable if there were two translocations (n=2).

to differences in gene *arrangements.*" Incompatibilities of this "chromosomal" type were of a structural nature (e.g., occurrence of an inversion of a segment of a chromosome of paternal origin, which might not have occurred on the homologous chromosome of maternal origin). Dobzhansky regarded chromosomal sterility as due to what may be called chromosomal "macromutations" (see chapter 12). These would be the same as the "quantitative," "intermolecular," mutations of Darlington. However, unlike Darlington, he considered sterility associated with *functional* defects in individual gene products ("genic" sterility) as most likely to be responsible for evolutionarily-significant events.

Tan in 1935 showed that *small* inversions (still "macromutations") were more likely to disrupt pairing than large inversions. Subsequent work in microbial systems was to suggest that multiple incompatibilities in single bases ("micromutations") would suffice (see chapter 12).

SUMMARY

Studies up to the 1930s indicated that reproductive isolation (selection) occurs either because (i) male and female gametes fail to fuse (pre-zygotic isolation) due to some barrier to gamete transfer or to a failure of the fusion process itself, or (ii) male and female genomes, once fused as a zygote, do not "agree" to work with each other to allow development (hybrid inviability), or (iii) male and female genomes do not "agree" to work with each other to form gametes for the next generation.

The latter phenomenon (hybrid sterility) shows that an initial conjugation between male and female is but the beginning of series of interactions between genomes. Initially, the genomes occupy a nucleus and generate sets of gene products which either do, or do not, work effectively together. At this developmental stage, direct interactions between the genomes themselves may not be critical. However, finally the genomes have to interact directly in the process of meiosis which occurs exclusively in the gonads.

The latter interaction may occur many years after the initial conjugation event (in males often only shortly before the production of mature gametes), and may involve the association within one organism of "complementary factors" independently acquired by each parent. These factors may correspond to no conventional phenotype, but may be attached to the same "residue" or "base" as factors determining the conventional phenotype. The complementary factors might be "chromosomal," having been generated by "macromutations," and operating structurally, or "genic," having been generated (usually) by "micromutations," and operating functionally. Neither in the 1930s, nor in recent times (White 1978; King 1993), was it envisaged that "chromosomal" factors could be generated by micromutations.

8 Why Sex?

> "I really do not understand the role of sex ... in evolution. At least I can claim, on the basis of the conflicting views in the recent literature, the consolation of abundant company."
>
> George C. Williams (1975 ch.13, 169)

Imagine a world without males in which each woman, on average, was able to produce two offspring per generation asexually. Both of these would be female. In the first generation there would be two women. In the second generation there would be four women. In the third generation eight women, and so on. With a similar limit to offspring number, in a sexual world a woman would be likely to produce one male and one female per generation. Only the latter, on being fertilized by a male, would produce further offspring. In the first generation there would be two individuals, male and female. In the second generation, there would still be two, and so on to the third generation, etc.. Since the winners in the biological struggle for existence are organisms which leave the most offspring, sex would seem very disadvantageous. Thus, Delboeuf (1877) noted:

> At first glance, one would be inclined to conclude that the most favourable condition for the continuation of the species would be that in which the individuals of the species reproduce themselves by fission or, at least, that they are perfect hermaphrodites [can fertilize themselves]. However, actually perfect hermaphrodism is the exception and sexuality is the rule. By virtue of which law did the separations of the sexes become predominant in nature, when, contrarily, everything seemed to be opposing its expansion? How did seemingly advantaged species [breeding asexually] cede their positions to other seemingly most disadvantaged species [breeding sexually]? Above all, this is where the principle of Natural Selection fails.

Weismann (1893 ch.14, 432) pointed out that, by allowing recombination, sexual reproduction would allow the uniting within one individual of diverse positive adaptations. However, once united, recombination can also separate positive adaptations. Asexual reproduction can be viewed as just an extreme form of inbreeding. Thus, if we can understand why inbreeding (the breeding of "like" with "like") might be disadvantageous, or conversely, why outbreeding is advantageous, we might be able to understand the adaptive advantages of sex.

INBREEDING AND OUTBREEDING

Among humans we recognize many causes of clinical infertility. Sometimes the "fault" lies with one partner, who, like a mule, will also not be able to reproduce with other partners. However, sometimes there is a *specific* incompatibility between members of a particular pair. If each member finds a new partner, the unions may be productive. Thus, Darwin (1875 ch.18, 145) noted: "It is by no means rare to find certain males and females which will not breed together, though both are known to be perfectly fertile with other males and females [of the same species]. ... The cause apparently lies in an *innate sexual incompatibility* of the pair which are matched."

In some cases the latter may be explained as an effect of *close* relatedness between the couples concerned. Thus Darwin (1859 ch.4, 96-7) writes:

> I have collected so large a body of facts, showing, in accordance with the almost universal belief of breeders, that with animals and plants a cross between different varieties, or between individuals of the same variety but of another strain, gives vigour and fertility to the offspring; and on the other hand, that *close* [Darwin's italics] interbreeding diminished vigour and fertility; these facts alone incline me to believe that it is a general law of nature (utterly ignorant though we be of the meaning of the law) that no organic being self-fertilizes itself for an eternity of generations; but that a cross with another individual is occasionally, *perhaps at very long intervals*, indispensable.

Darwin is here not just speaking about plants and animals considered high on the evolutionary scale. In a letter to Lyell (1861) he wrote (Darwin and Seward 1903 ch.3, 190): "I should certainly conclude that all sexuality has descended from one prototype. Do not underrate the degree of lowness of organization in which sexuality occurs – viz., in *Hydra* and still lower in some of the one-celled confervae which 'conjugate,' [and] which good judges ... believe is the simplest form of true sexual conjugation. But the whole case is a mystery."

An extreme example of inbreeding is the self-pollination which may occur in some plants. which Darwin (1859 ch.8, 249) notes: "would be injurious to their fertility." Thus, we can recognize at least three degrees of sexual compatibility with respect to the production of offspring: (i) When there is a high degree of genetic relatedness there is decreased fertility, and any offspring produced may be lacking in vigour. (ii) When there is a small decrease in genetic relatedness (but the union is between members of the same species), fertility increases and the offspring are vigorous ("hybrid vigour," "positive heterosis," or "overdominance"). (iii) When the genetic relatedness is low, fertility decreases (hybrid sterility or inviability), and this can serve as a definition of species (see chapter 3). Indeed, in one of his last books Darwin (1878 ch.12, 459-60) reiterated:

> It is an extraordinary fact, that, with many species, flowers fertilized with their *own* pollen are either absolutely or to some degree sterile; ... if fertilized with pollen from *another* individual [of the same variety] or [from a member of another] variety of the *same species*, they are fully fertile; but if [fertilization is attempted] with pollen from a *distant species* they are sterile in all possible degrees, until utter sterility is reached. Thus we have a long series with absolute sterility at the two ends; at one end due to the sexual elements *not having been differentiated* [from each other], and at the other end [due] to their having been differentiated *in too great a degree*, or in some *peculiar* manner.

Hence, if we were to plot the fertility of the cross of an average member of a species with other members of the *same* species, as a function of the degree of genetic relatedness, we might obtain a curve as shown in Figure 8.1. As relatedness decreases, fertility increases to a peak and then decreases. Vigour usually increases progressively as genetic relatedness decreases, even when fertility begins to decline towards the zone of full hybrid sterility. Darwin (1875 ch.17, 112) remarked: "For it deserves special attention that mongrel animals and plants, which are so far from being sterile that their fertility is often actually augmented, have ... their size, hardiness, and constitutional vigour generally increased [compared with the pure lines]. It is not a little remarkable that an accession of vigour and size should thus arise under the opposite contingencies of increased and diminished fertility."

We should recall (chapter 2) that, historically, the term "hybrid" has been used to describe the product of a cross between two variant organisms usually with marked anatomical differences. However, it is more convenient to regard the product of a cross between *any* genetically different organisms (be the difference only in one DNA base) as a hybrid. Thus, you are the hybrid of your parents. We discuss in chapter 12 the

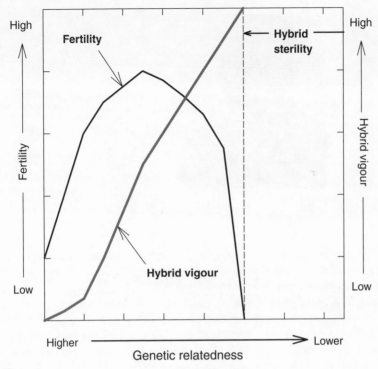

Figure 8.1
Relationship between the degree of genetic relatedness between two organisms, their fertility when crossed, and the vigour of the hybrids so-produced.

idea that the decrease in fertility as genetic relatedness decreases (sometimes called "negative heterosis" or "underdominance") might reflect the presence of incipient species within the major species.

WHY ARE HYBRIDS VIGOROUS?

By coming together as a hybrid, two genomes acquire the potential both to complement, and to repair, each other's defects. Regarding sexual conjugation in protozoa, Montgomery (1902) noted:

> From the studies of R. Hertwig and Maupas on Infusoria, it appears probable that conjugation or fertilization is essentially a process of rejuvenation: cells may divide and reproduce for a number of generations asexually, but there comes a period when the cellular vitality diminishes, so that no further reproduction is possible except after rejuvenation by conjugation with another cell. When thus rejuvenated by admixture of

Figure 8.2
Öjvind Winge (1886-1964), circa 1917.

substances from the other conjoint, the cell starts upon a new period of generation. ... From this standpoint the conjugation of the chromosomes in the synapsis stage may be considered the final step in the process of conjugation of the germ cells. It is a process that effects the rejuvenation of the chromosomes; such rejuvenation could not be produced unless chromosomes of *different* parentage joined together, and there would be no apparent reason for chromosomes of *like* parentage to unite. ... Of course, it is not a true analogy to compare conjugating Infusoria (i.e., whole cells) with conjugating chromosomes (i.e., a portion of cells). But still it is very probable that the two chromosomes unite temporarily for the same reason that two Infusoria do, that is, for an interchange of substances.

Along similar lines, Öjvind Winge (1917; Figure 8.2) when considering chromosome pairing noted: "It is now reasonable to suppose that two individuals on fusing together would to a certain degree be able to replace what might be lacking in each ... to help out each other's shortcomings. An individual failing ... may be repaired in its subsequent fusion with another, so that the gametophytes [haploid gamete-forming cells] again formed by the double [diploid] organism leave the diplophase in better condition."
Concerning the mechanism of this, Winge first alluded to what we would now consider the encoding potential of the chromosomes:

The hereditary differences [in the chromosomes] are not themselves visible; it is not the morphological differences in the chromosomes, but variations of a far more delicate nature, which give rise to the external morphology [of the organism] ... "Predispositions" [which we may now equate with coding potential] in the chromosomes cannot of course be supposed to be present as morphologically demonstrable units, but must rest on some chemical conditions in the chromosomes as lie far beyond the range of our present knowledge.

He then went on to consider chromosome pairing:

> The importance of this pairing must, I imagine, be that the chromosomes, two by two, supply each other with missing chromatic parts, possible ides or pangenosomes [genes] ... in case such could have been lost during the ontogenesis [development] of the sporophyte [diploid individual] – and which are necessary if the organism (species) is fully to retain its disposition, i.e., remain genotypically unaltered. The chromosomes have then ... effected an interchange of substances, so that while each in the main still has its original composition, all are, none the less, affected by the temporary pairing in the points where they have individually suffered loss ... .

That the chromosome itself would retain its individuality was underscored: "I cannot admit that the individuality of the chromosomes should necessarily be lost, even though interchange between them may have taken place. ... The position may be illustrated ... by pointing out that two ships at sea, without losing their "individuality" might well exchange certain wares in order to complete their stores; the chromosomes can doubtless carry on a similar traffic during their conjugation in the synapsis stage."

Finally, Winge considered the implications of this for the deleterious effects of close inbreeding, and attempted to answer the objection – if sex is so beneficial, how come the dandelions are doing so well?

> That inbreeding can also lead to unfortunate results, as is proved by instances from both animal and vegetable kingdoms, corresponds extremely well with my theory, as the closer relationship between gametes will often naturally preclude the mutual replacement of loss in the chromosomes, which, owing to their common derivation, lack the same units. The ill effects of inbreeding should thus be due to retrogressive mutations ... .
>
> Many plants, however, continue to develop and flourish even when self-pollination constantly takes place through many generations, just as various sexually abnormal [asexual] individuals, of *Taraxacum* [dandelion] for instance, thrive excellently by purely vegetative propagation. ... I would point out ... that a clone or a definite biotype may possibly retain its qualities for a long time – *especially when selection takes place* – and it is perhaps only reasonable, that nature should exert a selection among the manifold "vegetative" individuals ... [e.g., the "unfit" are stringently excluded from the population; see Fig. 5.1a for this monotypic pattern of evolution].

We will permit Huxley (1888) to have the last word: "If the conclusion at which he [Darwin] ultimately arrived, that cross-fertilization is favourable to the fertility of the parent and the vigour of the offspring, is correct, then it follows that all those mechanisms which hinder self-fertilization and favour crossing must be advantageous in the struggle for existence." From this perspective, targets of selection should include adaptations which assist the evolution of mechanisms to detect when reproduction is likely to be successful, both in the general sense (i.e., time of mating), and in the individual sense (with whom to mate). One example of the latter is the detection of major histocompatibility (MHC) differences between individuals (Roser et al. 1991; Manning et al. 1992).

SUMMARY

Asexual reproduction can be viewed as an extreme case of inbreeding within a species, which when contrasted with outbreeding among individuals of the same species, is generally found to be disadvantageous. Thus, if we can understand why inbreeding is disadvantageous, we may understand why asexual reproduction is disadvantageous. Darwin held that "a cross with another individual is occasionally, *perhaps at very long intervals*, indispensable," and noted that this was likely to apply to even the lowest life forms. Winge speculated that the adaptive value of sex was that it allowed the pairing of homologous chromosomes for mutual error-correction.

PART TWO

The Species-Dependent Component of Base Composition

9 Molecular Biology

"In nature hybrid species are usually sterile, but in science the reverse is often true. Hybrid subjects are often astonishingly fertile, whereas if a scientific discipline remains too pure it usually wilts."

Frances Crick (1988 ch.14, 150)

Mendel is generally thought of as a biologist. Yet his early training in Vienna was in physics, which may explain his emphasis on the quantitation of biological phenomena (Roberts 1929; Iltis 1932; Olby 1985; Forsdyke 2001b). It is hardly surprising that, in a world much awed by Einstein's new concepts of space and time, Bateson (1913 ch.2, 41) should seek to recruit physicists to the species problem: "It is I fear a problem rather for the physicist than for the biologist. ... I suspect that when at length minds of first-rate analytical power are attracted to biological problems, some advance will be made of the kind which we are awaiting."

In the 1920s and 1930s a group of biologists, including many who were mathematically inclined, were so bold as to attempt a "modern synthesis" on the assumption that the "missing link" in Darwin's arguments would be found in Mendelian genetics (Provine 1971; Ayala and Fitch 1997). However, their approach, and that of their followers, was largely genetical, with mathematical and rhetorical overtones which sometimes tended to obscure rather than enlighten (Provine 1986, 1992; Orr 1996; Forsdyke 1999c).

A NEW APPROACH

The independent involvement of chemists and physicists promised a completely new approach. "To break down the problem into simpler problems," they looked for the smallest possible biological system to which the attribute "live" could be appended. Darwin had become aware of the great proliferative powers of micro-organisms through the studies of Lionel Beale and John Burdon Sanderson on cattle plague (rinderpest) in the 1860s (Romano 1993, 105-91; 1997). In 1875 (ch.27, 372-3) he noted:

> Nor does the extreme minuteness of the gemmules [genes], which can hardly differ much in nature from the lowest and simplest organisms [to the highest], render it improbable that they should grow and multiply. A great authority, Dr. Beale, says "that minute yeast cells are capable of throwing off buds or gemmules, much less than the 1/100000 of an inch in diameter"; and these he thinks are "capable of subdivision practically ad infinitum."
>
> A particle of small-pox matter, so simple as to be borne by the wind, must multiply itself many thousandfold in a person thus inoculated; and so [also is the case] with the contagious matter of scarlet fever. It has recently been ascertained that a minute portion of the mucous discharge from an animal affected with rinderpest, if placed in the blood of a healthy ox, increases so fast that in a short space of time the whole mass of blood, weighing many pounds, is infected, and every small particle of that blood contains enough poison to give, within less than forty-eight hours, the disease to another animal.

Darwin's point was reiterated in 1922 by Herman Muller when commenting on the discovery of viruses which infect bacteria (called bacteriophages by Felix d'Herelle):

> That two distinct kinds of substances – the d'Herelle substances and the genes – should both possess this most remarkable property of heritable variation or "mutability," each working by a totally different mechanism, is quite conceivable, considering the complexity of the protoplasm; yet it would seem a curious coincidence indeed. It would open up the possibility of two totally different kinds of life, working by different mechanisms. On the other hand, if these d'Herelle bodies were really genes, fundamentally like our chromosomal genes, they would give us an utterly new angle from which to attack the gene problem. They are filterable, to some extent isolatable, can be handled in test-tubes, and their properties, as shown by their effects on bacteria, can then be studied after treatment. It would be very rash to call these bodies genes, and yet at present we must confess that there is no distinction known between genes and them. Hence we cannot categorically deny that perhaps we may be able to grind genes in a mortar and cook them in a beaker after all. Must we geneticists become bacteriologists, physiological chemists, and physicists, simultaneously with being zoologists and botanists? Let us hope so.

Bacteriophages seemed just right. Physicist Max Delbrück (1949) noted:

The complexities of sexual reproduction and of recombination are not eliminated by going to this seemingly elementary level ... . When a bacterium is simultaneously infected with two similar but different virus particles, the progeny will contain recombinants in high proportions ... . [Thus] ... virus ... reproduction must involve manoeuvres analogous to those occurring in meiosis and conjugation of higher organisms, ... [and the investigator has] the ability to control and vary the experimental conditions under which reproduction occurs.

Indeed, Lederberg and Tatum (1946) showed that bacteria themselves, long considered to multiply only asexually, were capable of undergoing various types of sexual activity (e.g., conjugation), leading to recombination and the transfer of genetic markers from one bacterium to another.

The revolution which then occurred brought non-biological scientists, calling themselves "molecular biologists," to the fore (Olby 1974). "The species problem" which so obsessed the biologists was not high on the research agenda of most molecular biologists. Biologists and molecular biologists appeared as two solitudes, two "islands of near conformity surrounded by interdisciplinary oceans of ignorance" (Ziman 1996). Unmoved by the successes of the molecular biologists, many biologists continued to address the problem using complex multicellular organisms such as the fruitfly.

SPECIES PROBLEM ADDRESSED

However, although most molecular biologists appeared unaware of the species problem per se, it was being addressed indirectly in their researches. They were showing that the genetic material, capable of varying yet carrying hereditary information through the generations, was DNA; that it consisted of four chemical building blocks (bases; A, C, G, T) which were present in definite proportions (e.g., Chargaff's first parity rule; A%=T%, C%=G%); and that DNA had a "duplex" double helical structure with either As on one strand pairing with Ts on the other strand, or Cs on one strand pairing with Gs on the other strand (Watson and Crick 1953).

Furthermore, they were intrigued by the mechanisms by which DNA molecules would break and recombine with other DNA molecules. Discovering how to carry this out in the test tube led to a revolution in biotechnology. Novel methods of rapid DNA sequencing transformed biochemistry laboratories which began to spew data so rapidly that only computers could handle them. The new science of "bioinformatics" emerged. Biology seemed to be becoming yet another branch of information science, a discipline which had been pioneered in the 1940s by workers at the Bell Telephone Laboratories (Shannon 1947; Khinchin 1957; Hamming 1980).

The application of some of the discoveries of the molecular biologists to the problem of speciation is the subject of this chapter. At the centre of the problem of reproductive isolation leading to speciation are variations in DNA sequences (accepted mutations). Each of our cells contains a set of chromosome pairs (homologs), one member of each pair being inherited from father (paternal set) and the other being inherited from mother (maternal set). If the sequence of a particular region of DNA in a chromosome is identical in paternal and maternal homologs then you are "homozygous" for that region. If the sequences differ, then you are "heterozygous" for that region. It should be noted that organisms of the same species which are heterozygous for certain regions reproduce successfully. Heterozygosity per se (often very small, localized, sequence differences), does not prevent reproduction. Indeed, heterozygosity at the DNA level may not be reflected in heterozygosity at the protein level (see the example sequences in chapter 1). Thus (and here comes a mouthful) although genotypically heterozygous, an organism may be phenotypically homozygous.

Much progress in recent years derives from observations on microorganisms (bacteria and yeast). Key proteins likely to be involved in speciation events are widely detected in living organisms (Hunter et al. 1996; Baker et al. 1996). Thus, it seems possible that the results from microorganisms are of general applicability. Further support for this view derives from bioinformatic studies carried out on the DNA of a wide range of species (Forsdyke 1995c-f).

RECOMBINATION BARRIERS IN BACTERIA

Although mainly reproducing asexually, the discovery of conjugation in bacteria provided an opportunity for investigating speciation at this more elementary level. *Escherichia coli* and *Salmonella typhimurium* are two closely related bacterial species. Since much of their DNA has been sequenced the degree of genetic similarity can be quantified; that is, the degree of sequence similarity can be measured. Assuming that the two species have arisen from a common ancestor, it can be shown that the two genomes are about 16% diverged. In simple terms, relative to the sequence of the ancestor, *E. coli* might have diverged 8% and *S. typhi* might have diverged 8%. Actually it would have been more complicated than this, since both organisms might have undergone some common changes (accepted mutations) relative to the ancestor, and there is both a forward mutation rate (away from the ancestral type), and a backward mutation rate (reverse of the forward mutation).

Mathematically-inclined biologists spend much time on such matters, but here we will not hesitate to simplify. Note the reference to "accepted" mutations. Many mutations are likely to be corrected, perhaps within seconds of their occurrence. What mutations are detected depends on the assay being used. In the literature, the word "mutation," unless qualified, usually means "accepted mutation." A "mutant organism" is an organism whose DNA has changed (mutated) relative to the control ("wild type") organism.

Members of two bacterial species growing together in the same environment may preferentially exchange their DNA with members of the *same* species (*intra*-species exchange). Thus there is reproductive isolation. However, there is a low rate of *inter*-species exchange (Baron et al. 1968). This provides the opportunity to isolate mutant bacteria which allow a greater rate of inter-species exchange (i.e., the barrier between species is decreased). In such mutant bacteria it would be expected that cellular components involved in some aspect of the exchange process would have altered (i.e., the DNA encoding those components would have mutated). Among these might be components responsible for critical self/not-self discrimination tasks; a bacterium can "ask" is this mating partner a member of my species ("self"), or of another species ("not-self")? Discrimination might be at the level of DNA transfer, and analogous to prezygotic exclusion. Alternatively, discrimination might affect DNA within the recipient bacterium (analogous to post-zygotic exclusion).

Such studies led to the recognition of a system of proteins already known to be involved in the repair of replication errors (mismatches) in DNA (MutH, MutL, MutS; Radman and Wagner 1993; Radman, Wagner and Kricker 1993). *E. coli* and *S. typhi* more readily exchanged their DNA when components of this system are damaged (Rayssiguier et al. 1989). Discrimination is at the level of DNA, and is so sensitive as to detect sequence divergence at the level of individual bases. Although the precise molecular mechanism is not known, it appears that at some stage potentially recombining DNA molecules are aligned with each other prior to the recombination event. If there is not an adequate level of similarity, then the process is aborted, and recombination is prevented. However, even though mismatch detecting proteins are defective, *inter*species recombination is still *much less* than *intra*species recombination (Zahrt and Maloy 1997). This suggests that the mismatch system can be an adjunct, but not a major, barrier to recombination. Indeed, the mismatch system does not impede recombination in two other groups of bacteria – Streptococcae (Humbert et al. 1995) and Bacillae (Majewski and Cohen 1998). In the latter case the initial alignment of chromosomes is still defective (termed "heteroduplex resistance").

RECOMBINATION BARRIERS IN YEAST

The mismatch recognition process appears to be universal since similar systems of "anti-recombination" proteins are present in yeast and in humans (Baker et al. 1996). Crosses between two closely related groups of yeast (11%-20% sequence divergence) are impaired (hence the groups are defined as species; Naumov 1987). That the anti-recombination system acts during meiosis is evident since the rare viable offspring from hybrids between members of the two yeast species often contain unusual numbers of chromosomes (aneuploidy). This implies misalignment of homologous chromosomes during meiosis so that there is not correct apposition of the parts (centromeres) required for equal partitioning (segregation) of chromosomes among daughter cells. The rare viable offspring also show lower frequencies of exchange between the parental DNA molecules than usual (impaired recombination). Mutations of proteins of the mismatch detecting system decrease the aneuploidy and increase the frequency of DNA exchange. However, the levels are *not* restored to those normally observed in crosses between members of the same yeast species (Hunter et al. 1996). Thus, in yeast as in bacteria, other factors may be involved in maintaining the recombination barrier. Recently interest has turned to early "stem-loop" recombination models which suggest another way "heteroduplex resistance" due to mismatches can contribute to the barrier.

STEM-LOOPS INITIATE THE HOMOLOGY
SEARCH

Critical to the acceptance of early stem-loop recombination models are recent observations suggesting that pairing of homologous chromosomes for recombination does not require the prior formation of a structure known as the synaptonemal complex. Furthermore, the homology search *precedes* strand-breakage, rather than the converse as proposed in a popular model (Szostak et al. 1983). Chromosomal pairing is likely to be the result of a "paranemic" DNA sequence-based homology search, which does not require an initial physical breakage of DNA strands (Hawley and Arbel 1993; Xu and Kleckner 1995; Rocco and Nicolas 1996; Kleckner 1997).

The story can be traced back to Muller's observation (1922): "It is evident that the very same forces which cause genes to grow [replicate] should also cause like genes to attract each other [when chromosomes pair at meiosis], ... If the two phenomena are thus dependent on a common principle in the make-up of the gene, progress made in the study of one of them should help in the solution of the other."

In 1941 he further predicted that "the gene ... is at least bilaterally complementary in its pairing and duplicational properties." A decade later James Watson (who had studied with Muller) and Francis Crick presented their double helix model for DNA. Shortly thereafter Muller set his students an essay "How does the Watson–Crick model account for synapsis?" (Carlson 1981 ch.27, 390). Crick took up the challenge in 1971 with his "unpairing postulate," which required that base-paired strands in duplex DNA would unpair to expose free bases. The exposed single-stranded regions would then pair with complementary single-stranded regions extruded from the duplex DNA of the homologous chromosome. This would initiate the recombination process (Figure 9.1).

However, single-stranded nucleic acids can adopt cruciform, stem-loop structures (Gierer 1966). In 1972 the model was modified by Sobell, who proposed that the single-stranded regions would be extruded as

Main axis of paternal chromosome

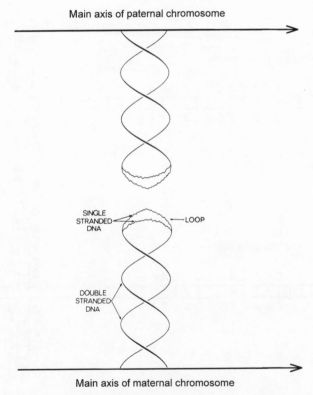

Main axis of maternal chromosome

Figure 9.1
Crick's unpairing hypothesis. Duplex DNA is extruded like a twisted hairpin at right angles to the main chromosome axis. Unpairing occurs at the tips of the hairpins, so that bases in single strands are free to pair with complementary bases.

stem-loops. The stems would be formed by intra-strand Watson–Crick base pairing and single-stranded loops at the tips of the stems would be available to initiate pairing between homologs. In Figure 9.2 the base triplet GAG at the tip of a paternal loop would pair with CTC at the tip of a maternal loop (Wagner and Radman 1975; Doyle 1978).

The first evidence for the model came from an unexpected source. In a certain micro-organism (a bacterial plasmid) the initiation of DNA synthesis is controlled by an RNA molecule transcribed from one strand of the DNA duplex. This RNA is regulated by an "antisense" RNA transcribed from the complementary DNA strand. Tomizawa (1984) made detailed kinetic studies of the process by which the antisense RNA reacted reversibly with the complementary "sense" RNA derived from the other DNA strand. Since these molecules were copies of the sequences of

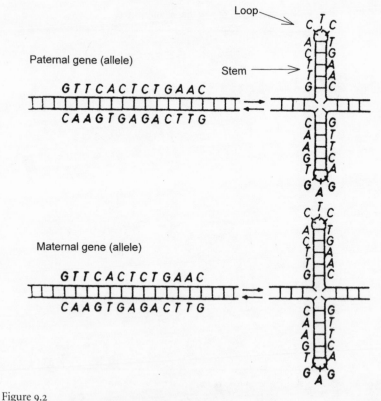

Figure 9.2
Pairing between complementary bases in loops. Provided bases are appropriately positioned to form stems, duplex DNA can extrude stem-loop structures. Intra-strand pairing initiates between complementary bases in the loops.

complementary strands of duplex DNA, studying the interaction of the RNAs might provide clues on how the single strands of the corresponding DNA molecules might recognize each other when "unpaired" for recombination.

Tomizawa concluded that the two RNA molecules first interacted through the loops at the tips of stem-loop structures. If this "kissing" were successful (i.e., precise sequence complementarity was found), then the interaction would be consummated by extending the pairing of sense and antisense strands. The pairing disrupted the stem-loop structure of the sense transcript and inhibited its role in initiating DNA synthesis (Figure 9.3). In this case the purpose of pairing was not to promote recombination, however the mechanism was consistent with the above stem-loop DNA recombination models. If sequence complementary was not found during "kissing" interactions between the loops extruded from homologous DNA duplexes (heteroduplex resistance), then recombination would not be initiated.

"Kissing" is a term used infrequently by biochemists when describing reactions between molecules. Tomizawa's choice of the term was precise in that he wished to convey the sense of an exploratory interaction which could either progress, or be reversed. He may also have had an intuition that the exploratory events observed between potentially uniting RNA molecules were of reproductive significance (Eguchi et al. 1991).

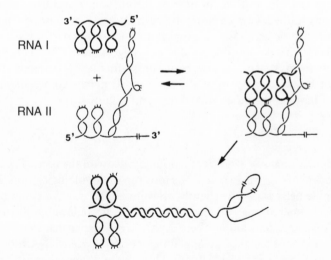

Figure 9.3
Tomizawa's loop–loop "kissing" model for the initiation of hybridization between two single-stranded nucleic acids. If the initial pairing is sufficiently stable, stems unfold and the pairing propagates to generate a duplex ("consummation").

IMPLICATIONS FOR BASE COMPOSITION AND ORDER

The stem-loop models had important implications for DNA composition, since to be able to form stems, *single-stranded* DNA should have complementary bases appropriately located (Figure 9.2). Thus single-stranded DNA should tend to obey the same parity rule as duplex DNA (A%=T%; G%=C%). Indeed, to a close approximation single-stranded DNA does obey the rule ("Chargaff's second parity rule"; Chargaff 1979; Prabhu 1993; Forsdyke 1995e, 2001f; Forsdyke and Mortimer 2000).

When extensive sequences of DNA became available in the early 1990s, it was possible to examine more precisely the distribution of stem-loop potential in single strands of DNA. The argument was that if recombination is evolutionarily advantageous then mutations in DNA which favoured recombination would tend to be accepted. Two types of favourable accepted mutation were envisaged: mutations of the genes encoding the enzymes which mediate recombination, and mutations which improve the ability of DNA to be a target of the enzymes (i.e., to act as a recombination substrate). The former mutations would have been localized to the regions of the genes. The latter mutations would have been dispersed. There would have been a genome-wide evolutionary pressure on primary sequences promoting the acceptance of such mutations. Indeed, highly significant stem-loop potential is widely dispersed in DNA from different human chromosomes and from all other organisms studied. Furthermore, the *order* of bases, not just which bases are present (base composition), makes an important contribution to the potential (Forsdyke 1995c-f; Seffens and Digby 1999). This is expected since, to form a stem, precise base complementarity is required. The order of bases in one arm of the stem must match (complement) the order of bases in the other arm of the stem (Figure 9.2).

SUPERCOILING

The popular image of DNA is of a double helix with the bases, like rungs of a ladder, linking the two helices. However, under biological conditions the double helix is usually slightly unwound. One way the strain of this "negative supercoiling" on the molecule can be relieved is to extrude single-stranded stem-loops. There is growing evidence that this unwinding is important for the "kissing" interaction between DNA molecules (Wong et al. 1998; Strick et al. 1998). Thus, if the present thesis is correct, the enzymes affecting supercoiling (topoisomerases) are likely to be among those playing a key role in meiosis (Smith and Nicolas 1998). Supercoiling is transmitted along the molecule from the region of topoisomerase action, and unpaired single-stranded regions should be extruded suddenly depending on the degree of supercoiling and local sequence characteristics. Thus for the initiation of a homology search required for meiotic pairing, two DNA molecules might need to be similar both in their degrees of supercoiling, and in their base compositions.

SUMMARY

The increasing involvement of physicists and chemists created a revolution in biology. Developments in biotechnology and bioinformatics led to studies of barriers to recombination between species of bacteria and between species of yeast. Mutations in the mismatch recognition system are insufficient to bring recombination between members of different species to the same level as recombination between members of the same species. Another anti-recombination activity acts at the level of the initial homology search. Past emphasis of the role of the synaptosomal complex in chromosome pairing has now given way to a stem-loop-based homology search model, which can be traced back to theoretical work of Crick in the 1970s. Predictions of this stem-loop "kissing" model are supported by bioinformatic analyses of base composition and of the distribution of stem-loop potential.

10 Primary and Secondary Levels of Information

> "It deserves especial notice that the most important objections [to my theory] relate to questions on which we are confessedly ignorant; nor do we know how ignorant we are."
>
> Charles Darwin (1859 ch.14, 466)

Genome sequencing projects reveal the number of genes in an organism, and the proportion of those genes which correspond to known functions. At the time of this writing (circa 1998) some small genomes have been entirely sequenced. About half the genes correspond to known functions. For the first time it appears we have some measure of our ignorance. However, genomes convey more than just genic information. We now return to the parallels between the evolution of languages and of species which began in chapter 1, and show that the words of a "pore flahr gel" can further dispel our ignorance.

Eliza Doolittle appeared to be reproductively isolated from Freddy Eynsford-Hill largely because of her accent. Although they could, perhaps with some difficulty, understand each other's primary information, the secondary information (especially Eliza's dropped Hs) seemed to constitute a barrier. Shaw tells us that Professor Higgins performed the appropriate experiment. He demonstrated that if the linguistic barrier were removed, then the reproductive barrier would also be removed; thus, Eliza and Freddy lived happily ever after (Shaw 1913).

It is important to note that initially the cockney accent was complete; there was no part of Eliza's speech which was not under its influence. With Professor Higgins's tuition the influence grew less, so that there was a stage at which sections of her speech were non-cockney, while others remained cockney. Finally, her speech became uniformly non-cockney. It should also be noted that once the marriage knot had been tied (i.e., the barrier to reproduction had formally been removed), there was a good chance that, even if her accent subsequently regressed, the marriage (reproduction) would continue. Thus the initial secondary information (a reproductive barrier) needed only to be removed temporarily until a more substantial (legal) way of removing the barrier had been substituted. Conversely, if there were a divorce (legal re-establishment of a reproductive barrier), then the secondary information (linguistic barrier to reproduction) might again become a factor in any alliances Eliza might contemplate.

TRANSMISSION AT DIFFERENT WAVELENGTHS

Another useful analogy is between the evolution of the radio system and the evolution of species. At the beginning of the twentieth century the first radio transmitter came into operation. A particular wave-band was chosen and messages were broadcast. The second radio transmitter to arise did not want to interfere with messages from the first. There were three alternatives. (i) The first and second transmitters could transmit in the same geographical area ("sympatrically") but would agree to transmit at different times (temporal isolation). (ii) The second transmitter could operate in a distant geographic area ("allopatrically") beyond the range of the first transmitter (geographical isolation). (iii) The two transmitters could operate at different wavelengths (intrinsic isolation). As the number of transmitters increased the third alternative became the most feasible. Thus different transmitters often broadcast both synchronously and within the same geographic area (Figure 10.1).

Operating on a second wavelength did not disadvantage the second transmitter. A message on one wavelength sounded much the same as on another wavelength. Thus the message was essentially wavelength-independent. This component of the radio signal can be regarded as the *primary information*. Information on wave-length is another (intrinsic) component of the signal and can be regarded as the *secondary information*. Again, it should be noted that this secondary information is widespread, affecting all parts of the primary message, and serves to identify (isolate) pieces of primary information arising from different transmitters. When two pieces of non-identical primary information are broadcast on the *same* wavelength the result is not pleasing to the ear. This indicates that "hybridization" of the two messages has not been successful, and pressures the operators of the transmitters to adjust at least one of the wavelengths.

THE SPECIES-DEPENDENT COMPONENT OF BASE COMPOSITION

DNA can also be considered to have primary and secondary information components. Table 10.1 shows three columns of data adapted from a 1952 study by Wyatt on the base composition of the duplex DNA of various

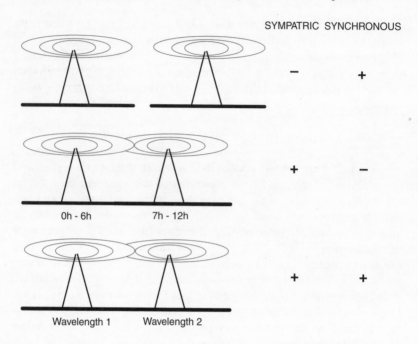

Figure 10.1
Three alternatives for avoiding interference between radio transmissions.

insect virus species. The species are classified according to their coat type (polyhedral or capsulated), and to their host type. The first two columns demonstrate the famous "Chargaff ratios" (Chargaff's first parity rule). Since A%=T%, then if the number of As is divided by the number of Ts, the result is about one. Similarly, since G%=C%, then the result of division is again about one. The data in these columns are species-invariant and led to the introduction of the general model for DNA by Watson and Crick (1953). Their model of DNA as a double helix offered explanations for many problems in biology. The "primary information" in DNA, such as information coding for the sequence of a protein, might be a characteristic of the sequence of the four bases. Individual DNA strands might act as templates permitting accurate replication of genetic information prior to cell division. One strand might be used as a template to repair damage in the other.

The great success of the double helix model may have distracted from the data in Wyatt's third column. This confirms Chargaff's suggestion that (C+G)% is a *species* characteristic. The ratio of the bases which engaged in strong Watson–Crick interactions (C+G), to the total bases

Table 10.1
Relative proportions of bases in insect virus DNAs

Virus type	Virus host	A/T	G/C	(C+G)%
Polyhedral	P. dispar	1.06	1.08	58.5
	L.monocha	1.03	1.08	51.5
	C. fumiferana	1.03	1.09	51.3
	P. seriata	1.04	1.05	47.6
	M. americanum	1.04	1.05	42.7
	B.mori	1.04	1.11	42.7
	C. P. eurytheme	1.08	1.11	42.5
	N. sortifer	1.07	1.09	37.4
Capsule	C. murinana	1.05	1.11	37.4
	C. fumiferana	1.01	1.12	34.8

(A+C+G+T), is constant for a particular virus species. This species constancy tends to be a characteristic of the entire genome, at least in the case of viruses and bacteria (Sueoka 1961b; Muto and Osawa 1987). Thus if one measures the sum of the Cs and Gs in a segment of 1000 bases in one part of the genome, one arrives at a certain value; this value is quite close to the value obtained from a 1000-base segment in another part of the same genome. The (C+G)% is *secondary information* distributed relatively uniformly in DNA. The genomes of organisms higher on the evolutionary scale tend to have their secondary information in large sectors (Bernardi 1993). Possible reasons for this will be discussed in chapter 11.

(C+G)% DIFFERENCES CAN CORRELATE
WITH PHYLOGENETIC DIFFERENCES

Wyatt and most other researchers in the 1950s and early 1960s were concerned with how DNA encodes its primary information, particularly proteins. Nevertheless, from studies of various strains of micro-organisms, Noboru Sueoka (1961b) speculated on evolutionary (phylogenetic) relationships based on (C+G)% differences (referred to as "per cent GC"): "An investigation of mean GC contents of twelve strains of *Tetrahymena pyriformis* ... showed that DNAs of these strains cluster in a rather narrow range, between 23 and 31 mole per cent GC. Moreover, the compositional heterogeneity of DNA molecules of each strain is relatively small. This provides further evidence that DNA base composition is a reflection of phylogenetic relationship."

He then went on to note that similarity of base composition was associated with successful mating, implying that *dissimilarity was associated with unsuccessful mating* (reproductive isolation). The differentiation of

DNA sequences which accompanied reproductive differentiation (mating ability) was characterized by *uniform* differentiations in base compositions:

> Furthermore, it is evident that those strains which [can] mate with one another (i.e., strains within the same "variety") have *similar* base compositions. ... If one compares the distribution of DNA molecules of *Tetrahymena* strains of different mean GC contents, it is clear that the difference in mean values is due to a rather *uniform difference of GC content* of individual molecules. In other words, assuming that strains of *Tetrahymena* have a common phylogenetic origin, when the GC content of DNA of a particular strain changes, all the molecules undergo increases or decreases of GC pairs in similar amounts. This result is consistent with the idea that the base composition is rather uniform not only among DNA molecules of an organism, but also with respect to different parts of a given molecule.

Further evidence on (C+G)% as an important species characteristic came from phylogenetic studies of retroviruses. Bronson and Anderson (1994) constructed genealogical trees to demonstrate the community of descent (phylogeny) of retroviruses from a common ancestor. Viruses whose sequences differed little from each other were placed closer on the tree than viruses whose sequences differed more. As a "control," they shuffled the sequences thus disrupting the primary information in the natural sequences, but retaining the secondary information (base composition). A unique genealogical tree was then constructed based on the information in the shuffled sequence.

This tree would have looked rather bizarre since it would have reflected the influence of the *particular* primary sequence which had been generated by shuffling, as well as base-composition. To eliminate the unique primary information component, they repeated the shuffling process a thousand times, and then took the *mean* alignment of the thousand individual alignments of the shuffled sequences which they had obtained. This alignment reflected *only* what was *common* in *all* the sequences which had been generated by shuffling, namely the base composition.

Intriguingly, in spite of the enormous information loss due to shuffling, a quite satisfactory genealogical tree of retroviruses could be constructed (Figure 10.2). Thus, a major component of the total information, *highly relevant to species differentiation*, seemed to have been retained. (C+G)% differences alone would suffice for construction of a phylogenetic tree. (C+G)% differences seemed to be linked in some way to retroviral evolution (phylogeny).

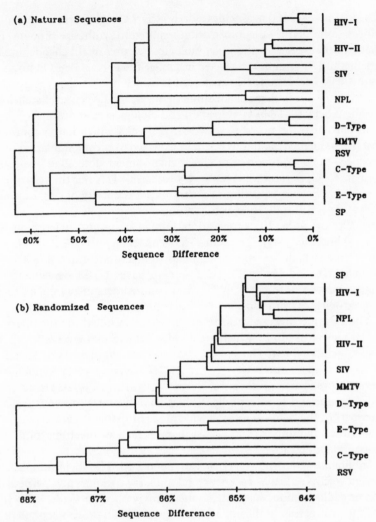

Figure 10.2
Phylogenetic trees of (a) natural and (b) randomized retroviral sequences. Similarities include the clustering of the C- and E-type sequences and of the lentivirus group (HIV-1, HIV-2, SIV, NPL).

VIRUSES IN A COMMON HOST CELL DIVERGE IN (C+G)%

In 1952 the double helical structure of DNA was not known. Wyatt knew that biologically different viruses (different host range) would be likely to encode different proteins to cope with different host environments. He was looking to (C+G)% for clues as to how DNA might carry its "primary information," namely the information required to synthesize proteins. He imagined that species with similar biological features might encode similar proteins and this might be reflected in similarities in (C+G)%. On inspecting his data (Table 10.1) he was forced to conclude that the (C+G)% "is identical in some biologically dissimilar viruses, and no general parallelism is evident between DNA composition and biological relationship."

In particular, he noted that two virus species which had a common host (*C. fumiferana*), and thus might be expected to encode similar proteins to cope with the common intracellular environment, differed dramatically in their (C+G) percentages (51.3% versus 38.4%). This was the very opposite of what he was expecting. Was this just a peculiarity of certain insect viruses, or was it indicative of something more general?

In support of Wyatt's observation, human immunodeficiency virus 1 (HIV-1) and the "E-type" human T-cell leukaemia virus 1 (HTLV-1), two retroviruses which have the potential to coinfect the same cell (a T lymphocyte; Yin and Hu 1997), are found to have (C+G) percentages of 42.6% and 53.2%, respectively (Bronson and Anderson 1994). Yet, these two retroviruses synthesize many similar proteins.

The extreme divergence in (C+G)% was also noted for members of the Herpes virus family which are divided into species groups (α, β, γ). These include the α-Herpes viruses (Herpes simplex 68%, Varicella zoster virus 46%), the β-Herpes viruses (HCMV, 57%, HHV6, 41%), and the γ-Herpes viruses (EBV, 60%, Herpes virus saimiri, 35%). Members of the α group tend to infect cells of the skin and central nervous system. Members of the β group infect a wide range of tissues. Members of the γ group infect lymphocytes and cause diseases such as infectious mononucleosis (glandular fever).

In linguistic terms, (C+G)% can be regarded as determining the "accent" of DNA. Just as cockneys tend to marry cockneys and perpetuate cockney information, so perhaps organisms with a particular (C+G)% "accent" will, at least at the initiation of the speciation process, tend to produce fertile offspring only with organisms of the same (C+G)% "accent." If this were so, the (C+G)% would be a dominant evolutionary force, constituting a "genome phenotype." Thus, Bernardi and Bernardi noted in 1986: "The organismal phenotype composes two components, the classical phenotype, corresponding to the gene products, and a genome phenotype, which is defined by [base] compositional constraints."

SUMMARY

Returning to the language theme begun in the first chapter, we consider the concept of two forms of information concurrently transmitted within one medium. The purpose of the second form is exclusionary, namely to act as a barrier preserving the purity of the first form. Helpful metaphors for this are *wavelength* in radio transmission of information, and *accent* in spoken information. In the case of DNA, the species-dependent component of base composition plays this secondary role. However, the barrier to recombination initially established by differences in this component may eventually be superseded by other barriers. Evidence for the initial barrier may then be hard to find, except in certain viruses which may not have advanced beyond the first barrier.

11 The Dominance of the Genome Phenotype

"According to the genome hypothesis each ... species has a 'system' or coding strategy for choosing among synonymous codons. This system or *dialect* is repeated in each gene of a genome and hence is a characteristic of the genome."

Richard Grantham et al. (1986)

In chapter 1 two sequences were shown, each encoding the same protein, but differing dramatically in (C+G)%. This was possible because the genetic code is a degenerate code, so that a particular amino acid in a protein can be encoded by one of a range of possible codons. Thus CG-rich codons or AT-rich (CG-poor) codons may be used to encode a particular amino acid.

The first amino acid in the sequences shown in chapter 1, phenylalanine, was coded for by TTT in the first sequence, and TTC in the second sequence. An accepted mutation of the codon TTT in the first sequence, so that it became TTC, would be a "synonymous mutation" in that it would not change the amino acid which was encoded. With respect to changing the sequence of the encoded protein the mutation would be "silent." Mutations which change codons so that a different amino acid is encoded are called "non-synonymous" or "non-silent" mutations. The change in an amino acid does not necessarily mean that the function of the corresponding protein will change, and even if that function does change, the change may be insufficient to affect the function of the organism. Thus, a "non-silent" mutation may be phenotypically silent.

Usage of alternative codons is not random. An organism doesn't "toss a coin" to decide which synonymous codon to use in the case of a particular amino acid. A widely held explanation of the divergence in (C+G)% between different species is that, since environmental selective pressures on organisms of a species act by way of the protein phenotype, as long as the right protein is made the sequence encoding that protein is free to vary depending on factors of minor evolutionary significance, such as species-specific "mutational biases" (Kimura 1989; Filipsky 1989; Suoeka 1995). The environment is held to dictate to the proteins, which, in turn, dictate to the genome. The genome, like Mary's little lamb, should follow the proteins faithfully, and there should be no need for an accommodation to genome (C+G)% by protein sequences.

On the other hand, if (C+G)% has an important role to play in evolution then there may be circumstances under which proteins are *forced* to follow (C+G)%. Redundancy of the code allows considerable flexibility in amino acid choice, but the flexibility is not absolute. There may be circumstances in which the genome "wants" *both* to have a particular (C+G)% *and* to encode a particular amino acid, but the two desires are in conflict. The needs of the genome may then dominate. Mary may have to follow the lamb. This has been found to apply in the case of bacteria (Sueoka 1961a), and other organisms including retroviruses (Bronson and Anderson 1994; Berkhout and Hemert 1994).

The genome (C+G)% is a major factor determining which codons are used (Nichols et al. 1981; Kagawa et al. 1984). Not only are proteins sometimes obliged to follow the genome (C+G)%, they may also be obliged to follow the genome's stem-loop potential. This genome dominance is the essence of Richard Grantham's "genome hypothesis."

GRANTHAM'S "GENOME HYPOTHESIS"

Among the first nucleic acids to be sequenced were those of small single-stranded RNA viruses, such as bacteriophage MS2. From knowledge of the genetic code it was then possible to derive the sequences of encoded proteins. It was also possible to investigate the most energetically favourable folding of the RNA into a secondary structure with Watson–Crick base pairing (e.g., Figure 9.2). It was found that there was a high degree of secondary structure (Salser 1970), and that in places the secondary structure function seemed to dominate the protein-encoding function. Ball (1973) concluded: "The amino acid sequence of MS2 ... has been subjected, during its evolution, to rearrangement *in the interests* of the secondary structure of the messenger RNA. This indicates that there is a pressure for some amino acid sequences to be selected according to criteria which are distinct from the structure and function of the protein they constitute."

Globin mRNA is another convenient RNA source for sequencing. Red blood corpuscles are highly specialized cells containing almost exclusively the proteins α-globin and β-globin, which form part of the oxygen-carrying haemoglobin molecule. Accordingly, the cells from which red blood corpuscles are derived ("reticulocytes") are actively synthesizing these globin molecules and are an excellent, highly enriched, source of the corresponding messenger RNAs (mRNAs). Following Ball (1973) and

Fitch (1974), Salser in 1978, when comparing the first globin mRNA sequences, again challenged the protein-centred view of sequence conservation during biological evolution:

> The conspicuous lack of silent [synonymous] base substitutions in the region [highly conserved between different globin protein sequences] ... argues that here it is the importance of the *nucleotide sequence* itself which is responsible for its conservation. It is, of course, novel to propose that an amino acid sequence of substantial size might be conserved because it was coded by a critical mRNA sequence rather than because the amino acid sequence was critical to the protein *per se*.

Salser went on to show that base substitutions predominate in mRNA *loop* regions, which are least likely to be involved in maintaining the configuration of the folded RNA structure. However, Salser's evolutionary perspective was still in terms of the fitness of individuals to meet the obvious day-to-day challenges of their environments. An RNA molecule might have a certain configuration, perhaps because it facilitated some regulatory event involving RNA–protein interactions. In some cases this is true, but it is also possible that nucleic acid structure is important not only at the RNA level, but also at the level of the DNA from which an RNA was transcribed. In this circumstance, it would be possible that the needs of DNA would predominate over the needs of mRNA and proteins. Their sequences would reflect an evolutionary compromise. Certain aspects of RNA and protein sequences would reflect the underlying needs of the encoding DNA, which are identified here as stem-loop potential (chapter 9) and (C+G)% (chapter 10).

In his "genome hypothesis" Grantham pointed out that *all* genes in a virus or bacterial genome tend to use the *same* subset of the 64 possible codons. The needs of the genome seem to dominate codon choice. The codon subsets of a virus and of its host are often different, even though they use the *same* translation machinery (Grantham et al. 1985). The same principle applies to sectored vertebrate genomes. Thus, mammalian α-globin and β-globin mRNAs share a common cytoplasm (within reticulocytes), yet use different codon subsets. The corresponding genes are in the high and low (C+G)% genomic sectors (isochores), respectively. It seems it is the *genomic* sector, by virtue of its (C+G)%, which is determining codon choice (rather than vice versa).

THE SELECTIONIST DILEMMA

Possible explanations of species differences in (C+G)% are at the heart of a debate between two schools. The "neutralists" propose that species-specific mutational biases result in mutations which are usually selectively neutral and do not provide phenotypic targets for discriminatory forces in the environment. Such "neutral" mutations are accepted into populations indiscriminately due to "random drift" (Kimura 1989; see chapter 5). The "selectionists" propose that some selective force drives species to adopt particular C+G percentages (Bernardi and Bernardi 1986). However, the neutralists have been able to point to examples of mutational biases (Cox and Yanofsky 1976), and seem readily to demolish the arguments of the selectionists.

For example, the bonds between a C and G base pair in DNA are less readily broken at high temperatures than the bonds between A and T. The selectionists argue that high (C+G)% genomes might have evolved in response to the selective pressure of temperature. Indeed, many contemporary organisms which survive at high temperatures do have CG-rich DNA (Kagawa et al. 1984). Yet, as neutralists point out, some organisms living at high temperatures have AT-rich DNA (Filipski 1989). *Certain* RNAs (e.g., tRNA, rRNA), whose predominant feature is *structure* rather than coding capacity, are found to have an increased content of C and G in micro-organisms living at high temperatures, but this is *not* found in the mRNAs of the same organisms, the latter RNAs reflecting the (C+G) content of the organisms' DNAs (Galtier and Lobry 1997). DNA seems to adapt to high temperatures, not by changing its base composition, but by associating with certain stabilizing molecules (polyamines; Oshima et al. 1990), and by relaxation of tortional forces (supercoiling; Friedman et al. 1995). Thus, genomic (C+G)% differences cannot be explained as an adaptive response to high temperature. The selectionist dilemma is that of finding a selective force which could have driven the evolution of species differences in C+G percentages.

ECOLOGICAL NICHE HYPOTHESIS FOR
EXTREME (C+G)% DIVERGENCE

Wide differences between two viruses in (C+G)% (chapter 10) would mean that, if they occupy the same host cell (thus appearing to use the *same* translation apparatus), they would differ widely in their choice of codons. This would allow them to synthesize similar proteins (converting their primary nucleic acid information into primary protein information), while retaining sequence diversity at the nucleic acid level (carrying both primary information which might be shared with another virus species, and secondary information which might be species specific). Some codons are more efficiently translated than others. However, the choice of codons would not appear to be related to the efficiency of translation, since both viruses would be presumed to require efficient translation, yet they differ widely in their choice of codons.

To explain the extreme (C+G)% differences, it was suggested that the intracellular environment could be considered to consist of a variety of "ecological niches" (Schachtel et al. 1991; Bronson and Anderson 1994). Two viruses which might simultaneously infect the same host cell would avoid competition with each other by each occupying a unique intracellular niche. A virus with a low (C+G)% would make different demands on the pool of precursors of C and G in its host (a metabolic ecological niche) than a virus with a high (C+G)%. Thus, the genomes of the ancestors of the two viruses would first have had a common (C+G)% percentage. Within a common host cell each virus species would have exerted a selective pressure on the other by competing for metabolites such as nucleic acid precursors. This would have constituted a powerful "C+G pressure" favouring the selection of mutants which deviate from the common (C+G)%.

The ad hoc ecological niche postulate was made because there did not seem to be an obvious adaptive reason for the extreme variation in (C+G)%. However, with the problem of speciation in mind, an alternative hypothesis would be that the selection pressure was the speciation process *itself*. In the absence of special mechanisms to prevent co-infection (analogous to pre-zygotic isolation), two species of virus coexisting synchronously and sympatrically within the same host cell would have had every opportunity to recombine. Thus, reproductive isolation would be lost and the viruses would mutually destroy each other as distinct species.

Since the viruses shared the challenges of a common environment it is likely that they would have some common proteins with similar sequences. If these sequences were also similar at the DNA level then recombination would be possible, since recombination between DNA molecules is favoured by sequence homology. Thus there would have been a strong selective pressure on the DNA of the two viral species to evolve in any manner which might prevent recombination (analogous to post-zygotic isolation operating at the level of meiosis). If differences in (C+G) percentages could *prevent* recombination, then reproductive isolation would be achieved.

VIRUSES ANALOGOUS TO GAMETES

Viruses are a special class of pathogenic agent in that they contain information for self-replication, but do not possess the resources to express and execute that information. For this they must enter a host cell, much as a small male gamete (sperm), containing information but very deficient in resources, must fuse with a female gamete (ovum). The latter contains abundant resources as well as information.

As discussed in chapter 4, a cross may be sterile (leading to reproductive isolation between members of the crossing species) because (i) the male gamete cannot reach or fuse with the female gamete (pre-zygotic exclusion), or (ii) the two, once fused, cannot "agree" to work together to allow the development of an adult organism (one form of post-zygotic exclusion), or (iii) the two cannot "agree" to work together in the gonad to allow the production of gametes for the next generation (hybrid sterility; another form of post-zygotic exclusion).

Three similar levels of reproductive isolation may be recognized among virus species. Two viruses may have a different host range, or, if they co-infect a multicellular host, they may target different cell types. If they do happen to target the same cell type, one virus may have evolved a mechanism to adapt the infected cell to prevent superinfection by another virus (the first one in "bolts the door"). These examples would all be analogous to pre-zygotic exclusion. If the latter is absent and two viruses can both access the same host cell, then, in the absence of some mechanism analogous to post-zygotic exclusion, recombination might occur. There would then be a strong selection pressure favouring evolution of the virus in a way which might prevent such recombination, thus allowing each virus species to retain its genetic individuality (remain part of a species).

(C+G)% COMPATIBITY NEEDED FOR SUCCESSFUL RECOMBINATION

If precise (C+G)% compatibility (equivalence of secondary information) were essential for successful recombination then there would be a selection pressure on the viruses to diverge in this "secondary information," while retaining the necessary "primary information" in their sequences. In the absence of the evolution of any other form of reproductive isolation between the two virus genomes, this evolutionary pressure would be *continuous*, driving the two (C+G) percentages far apart and *maintaining* the differences. Relief of the pressure would allow (C+G)% to become free to respond to weaker pressures, including mutational biases.

In the virus pairs discussed in chapter 10 (C+G) percentages have been driven and have *remained* far apart, suggesting that other characteristics of the viruses make evolution of some form of pre-zygotic exclusion difficult. The evolutionary strategy of the AIDS virus (HIV-1) appears to be that of maintaining a high mutation rate (by virtue of a highly error-prone DNA polymerase) in order to evade host immune defences (by providing a "constantly moving target"). Having mutated itself almost to oblivion, the virus rescues itself by actively recombining with other members of the "quasi species" population (all derived originally from one infecting virus particle). Although probably deriving from the same ancestral viral species, human T-cell leukaemia virus 1 (HTLV-1) has differentiated for a more passive strategy of long-term latency. However, if HTLV-1 were in the same T-cell host as HIV-1, then recombination to generate a less well-adapted hybrid virus would be possible (Yin and Hu 1997). Mutual destruction might result. Thus, if differences in (C+G)% could decrease recombination there would be a sustained pressure on (C+G)% so as to maintain and increase the difference.

HOW (C+G)% MIGHT AFFECT RECOMBINATION

Is there anything we know about the process of recombination which might be critically affected by (C+G)%? We have mentioned (chapter 9) the roles as anti-recombination agents of various proteins involved in mismatch detection, but there is no known link with (C+G)%. We have also presented evidence that the potential to form stem-loops is important for recombination. Here there is a link with (C+G)%.

The link derives from studies of various parameters affecting the computer folding of nucleic acids into stem-loop structures. Of a range of variables studied, by far the most important is the *product* of the quantities of C and G. This was first shown by Maizel and coworkers (Chen et al. 1990), and has been confirmed (Forsdyke 1998). The fact that it is not just their *sum* (C + G), but their *product* (C × G), gives the quantities of these bases great potential to influence the pattern of stem-loops which a duplex DNA molecule can extrude in single-stranded form for the loop–loop "kissing" interactions critical in the homology search preceding meiosis. Thus, we may conclude that (C+G)% can indeed be regarded as "secondary information," akin to wavelength in the transmission of radio messages. Different biological species can be considered to "broadcast" their DNAs at different (C+G)% "wavelengths."

Returning again to the somewhat imperfect radiowave metaphor (Figure 10.1), two transmitters which are close to each other (i.e., their transmitting ranges overlap) cannot broadcast simultaneously on the same wavelength. Similarly, two species which are biologically close to each other (i.e., pre-zygotic isolation may be imperfect), cannot "broadcast" on the same (C+G)% "wavelength" without interfering with each other. Two transmitters whose transmitting ranges do not overlap can broadcast at the same wavelength. Similarly, two biologic species which are reproductively isolated (through one or more pre-zygotic isolation factors), can "broadcast" their DNAs at the same (C+G)% "wavelength." If post-zygotic isolation precedes pre-zygotic isolation, then initially reproductive isolation would require different C+G percentages; when pre-zygotic isolation was achieved, the two C+G percentages could converge, since there would no longer be a selective pressure for (C+G)% divergence.

Thus in many modern species the process responsible for the *initiation* of reproductive isolation could have been *disguised* by subsequently developing pre-zygotic isolating factors which would ensure *maintenance* of reproductive isolation. This would then allow some *intragenomic* diversification of (C+G)%. In some cases the intragenomic diversification might serve the same purpose as interspecies genome diversification, namely the prevention of recombination. Once a prototypic globin gene had diversified ("paralogously") into α-globin and β-globin genes, migration and adaptation to regions differing in (C+G)% (isochores, which may be on different chromosomes) would militate against intragenomic recombination. Intragenomic diversification of (C+G)% may also have promoted the differentiation of sex chromosomes (see chapter 19).

PARENTAL GENOMES IN ALLOPOLYPLOIDS PRESERVE (C+G)%

Among the first studies of hybridization in plants were those of Joseph Kölreuter, who was greatly puzzled by the hybrid sterility of the "first botanical mule," the diploid product of a cross between certain species of *Nicotiana* (Roberts 1929 ch.2, 41-6). He resolved not to "break my head" on this "most complicated of knots." Two modern species, *Nicotiana sylvestris* and *Nicotiana tomentosiformis*, both with twelve pairs of chromosomes and differing in (C+G)%, are estimated to have diverged from a common ancestral species about seventy-five million years ago (Okamuro and Goldberg 1985). These are the parental species which about six million years ago gave rise to the allotetraploid *Nicotiana tabacum*, from which tobacco is manufactured. Since polyploidy "cures" hybrid sterility (see chapter 7), the hybrid with twenty-four pairs of chromosomes is perfectly fertile (Figure 7.1). For six million years the two parental genomes have occupied the same nucleus yet have retained their differences in (C+G)%, indicating minimum recombination between the genomes. This apparent genomic isolation has been attributed to "nuclear architecture" (Matassi et al. 1991), but may be a *consequence* of the differences in (C+G)%.

NEUTRALISM "TOO EASY"

In this chapter we have considered how differences in (C+G)% might influence recombination. Many studies have been carried out on the molecular basis of recombination and many valuable models have been advanced taking into account the results of such studies (Szostak et al.

1983; Holliday 1990). While further critical definition of the role of (C+G)% is needed, the view of the dominance of the genome phenotype presented here stands in strong contrast to the "neutralist" alternative (Kimura 1989; Sueoka 1995). As Grantham and his colleagues point out (1986): "It is just too easy to say most mutations are neutral." Indeed, it will be shown in chapter 15 how many polymorphic mutations, which have been considered neutral, might actually be adaptive.

SUMMARY

The need to convey two major forms of information provides an explanation for many puzzling features of DNA. These include the disparate codon choices of different species, the fact that the evolutionary "tune" is often called by nucleic acid rather than by protein, the apparent neutral nature of some mutations, and the wide divergence in the species-dependent component of base composition in the case of viruses with the potential to occupy a common host cell. New bioinformatic evidence suggests a mechanism by which differences in this component might prevent recombination by impeding a stem-loop-based homology search. In most modern species this initial post-zygotic isolating mechanism would have been substituted by subsequently developing pre-zygotic isolating mechanisms which would ensure maintenance of reproductive isolation.

12   Initiation of Speciation

> "At this point comes the inevitable question. What makes the character group split? ... With this question we come sharply to the edge of human knowledge .... We await our Pasteur."
>
> William Bateson (1904a, 257-8)

Modern species derive from common ancestors which are now extinct. At one time-point there was a set of interbreeding organisms which could be considered as one species. At a subsequent time-point that original species had diverged into two new species, A and B. Between these time-points the DNA sequences of the two groups changed in such a way that it became more advantageous for members of A to reproduce with members of A, and for members of B to reproduce with members of B. Given the great diversity of living organisms, one would expect there to be a variety of ways for members of two incipient species to diverge from each other. Some ways might involve initiation by way of pre-zygotic exclusion mechanisms, others by post-zygotic mechanisms. Our main interest is to consider the evolutionary advantages and initial molecular mechanisms which might be common to a *majority* of the processes which have led to reproductive isolation. We here consider post-zygotic incompatibilities at the level of meiosis, as discussed in preceding chapters, as providing the most likely *general* basis for *the* origin of species.

SEX AND RECOMBINATION REPAIR

Sexual reproduction provides an opportunity for DNA from two individuals to recombine. The new gene combinations which result may be advantageous (Weismann 1893 ch.14, 432; Zeyl and Bell 1997). However, recombination can disrupt as well as create favourable gene combinations. For this reason, among others, it is difficult to accept the generation of advantageous gene combinations as the main driving force favouring the evolution of sex. There has been much debate on the possible advantages of sexual reproduction compared with asexual reproduction (see chapter 8; Williams 1975 ch.1, 7-14; Maynard Smith 1989 ch.12, 237-54). We will here accept the postulate, recently reargued eloquently by Bernstein and Bernstein (1991; 1997), that recombination (sex) evolved primarily to correct DNA damage and mutations. Mutations are the source of all evolution since without changes in DNA there would be no variations upon which selective forces could act. The rate of mutation is usually such that repair is needed, but a population in which repair processes were too efficient would have a smaller range of variant organisms, so that the abilities to adapt to environmental challenges, and to speciate, might be decreased.

Mutations include changes in individual bases in DNA (substitutions, insertions or deletions), which may be considered as "micromutations," and larger changes affecting many bases, which may be considered as "macromutations." The latter terms as used here reflect the mutational processes as occurring at the DNA level, and not the phenotypic results of these processes (i.e., some refer to a very large difference in the appearance of an offspring relative to its parents as a "macromutation" or "sport"). There is no necessary correlation between the extent of a change at the DNA level and the corresponding change in phenotype. A micromutation might produce dramatic changes in phenotype, whereas a macromutation might cause no observable change.

Causes of mutations include DNA damage and replication errors. DNA damage may result from physical or chemical agents in the environment (e.g., certain types of radiation, and certain ingested chemicals which are not readily detoxified; Macdougal 1911; Muller 1928; Allen 1978 ch.4, 148), or may be an intrinsic part of the "cost" of running the metabolic machinery of the cell (e.g., highly reactive molecular species known as "free radicals" are generated within cells and may damage DNA).

Two cell strategies for dealing with the constant "breakdown" of its parts are *recycling* (so that molecules are degraded and then resynthesized from their component parts), and *repair*. The former strategy applies to four of the five major classes of macromolecules in living systems (lipids, carbohydrates, proteins and RNA). The latter strategy applies to the fifth class, DNA.

There are two general ways recombination might contribute to repair. On one hand, recombination would allow the collection together of deleterious mutations in one of the products of recombination, so that the mutations could subsequently be purged from the population by natural selection ("purifying selection"; Zeyl and Bell 1997). On the other hand, one genome can provide a template for correction of another. This is an active process not directly implicating natural selection. Organisms use recombination, among other methods, to repair (maintain the integrity of) their DNA. This was recognized in the phenomenon of "gene conversion" studied extensively in the bread mould, and is now recognized in a range of organisms including humans (Smithies and Powers 1986; Bernstein amd Bernstein 1991 ch.11, 241). It is unlikely that a DNA molecule in one member of a species will be damaged or mutated at the same site as in the homologous DNA molecule in another (not closely related) member of the same species. One DNA molecule thus has the potential to act as a template for repair of, or to replace a defect in, another.

Either a damaged base, or a mismatched base in a heteroduplex, can be dealt with. In the latter case, some mechanism to decide which is the correct strand is important. Sometimes strand-specific methylation or breakage is used to determine which strand to repair (Hare and Taylor 1985). Sometimes there might be comparison with another intact duplex (Forsdyke 1981). In the G2 phase of the cell cycle, diploid cells are temporarily tetraploid, so at this time there may be a better chance of "knowing" which strand to repair when a base in one strand of one duplex is mutated.

Intriguingly, following X-irradiation polyploid cells have less accepted mutations than euploid cells (Stadler 1929, 1932). This may be dismissed as indicating some "sheltering" due to excess copies of the gene in question. However, the possibility arises that the extra gene copies facilitate accurate error-correction (by providing more normal duplexes for comparison), rather than simply supply a redundancy of gene products. Thus there should be a selective pressure to maintain all gene copies intact and avoid the degeneration of some copies which might result in non-functional "pseudogenes." Indeed, pseudogenes are rare in polyploid species (Larhammer and Risinger 1994). Only under very special circumstances do extra gene copies degenerate and disappear (Navashin 1934; Du Pasquier et al. 1977; Reeder 1985). This is consistent with the idea that polyploid cells are larger than normal because there must be a balanced expression of all genes in order to sustain collective protein functions (see chapters 18 and 19).

MONITORING DEVIANCE FROM THE SPECIES NORM

Organisms of different species have different DNA sequences. Thus, to act as an accurate template for repair, DNA molecules must be from the same species. However, it is advantageous for organisms to avoid recombination with other members of their species whose DNA has deviated from the species norm (potential "incipient species"). The DNA of these members is no longer a reliable template for error detection and correction. Deviant members of a species can thus be seen as exerting a selection pressure for the evolution of mechanisms to monitor deviance from the species norm, and to prevent recombination if that deviance exceeds some unacceptable limit.

Ideally, the monitoring process would, in some way, first directly summate all the deviant aspects of a genome and then reject a genome which exceeded the limit. Thus, when you set forth seeking a mate, you should carry with you a DNA sequencer and a computer. The genomes of prospective mates would be sequenced in their entireties and compared with your own. Of course, much of the colour would disappear from a life lived this way!

In practice, many organisms rely on secondary physical or psychological characteristics, which might provide an indirect measure of the fitness of the DNA sequence within the intended mate. Similarly, *at the DNA level* the monitoring process could assess some characteristic which would provide an *indirect* measure of genome deviance, such as (C+G)%. In this case there should be some explanation why different species sometimes have very similar C+G percentages; this has been given in the preceding chapter, and will be elaborated below.

MULTIPLE STEPS IN THE INITIATION OF SPECIATION

Figure 12.1 shows the hypothetical (C+G)% distribution in two populations (*A, B*) which have successfully diverged on the basis of total genomic differences in (C+G)% and have undergone complete post-zygotic isolation. Each population has a bell-shaped distribution. Two genomes, one from each population, which happen to meet within a cell would not recombine because of the differences on (C+G)%. Hence, under the reproductive definition of species (chapter 3), *A* and *B* would be distinct species. The complete separation of the two bell curves would remain until some other form of isolation could supervene (e.g., pre-zygotic isolation such as the physical inability to conjugate). The latter would remove the pressure for (C+G)% divergence and the two (C+G) percentages would more easily respond to other pressures including any mutational biases. This might result in the two curves eventually overlapping and even merging.

The original pressure for (C+G)% divergence ("genotypic divergence") would have been the anti-recombination effect of such divergence. Anti-recombination would be most advantageous when it preserved beneficent phenotypic divergence. Thus, there are two interdependent types of mutational events. (C+G)% divergence ("genotypic" because the mutations involved will not necessarily influence the phenotype) will be most beneficial when there is phenotypic divergence (resulting from mutations which have affected the phenotype). On the other hand, phenotypic divergence *will not be sustained* unless there is genotypic divergence, resulting in reproductive isolation (Figure 12.1).

The possible merging of the two curves if pre-zygotic isolation became established would occur slowly, since it would be partly resisted from within each species. The average (C+G)% of the population would be the norm, the yardstick against which individual members of the species would compare themselves. This would be necessary because to be an accurate within-species template for recombinational repair, a genome should not have deviated far from the population norm, and (C+G)% would be an indicator of this. For each species recombination would be most frequent between individuals corresponding to the peak of the curve since these would be the most abundant types and so would be more likely to encounter each other for mating.

On the other hand, recombination would be infrequent between individuals with (C+G)% at the outer extremes of the curves. They would have deviated in (C+G)% from the species norm and would have the potential to become separate as "incipient species." Since they would be more likely to encounter potential mates with (C+G)% nearer the norm, such matings, if successful, would result in offspring with (C+G)% nearer the norm. On the other hand, the percentage of success (mating and

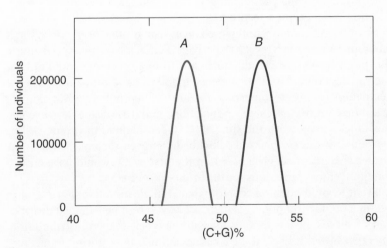

Figure 12.1
Two populations completely post-zygotically isolated on the basis of differences in (C+G)%. A rare member of species *A* at the extreme of the distribution (46% C+G) would show impaired fertility when crossed with an abundant member of species *A* (47.5% C+G), and even greater impaired fertility with a member at the other extreme of the distribution (49% C+G).

the production of fertile offspring) would be lower because of the extreme deviation in (C+G)% relative to that of most potential mates. On rare occasions, individuals with (C+G)% at the same extreme of the curves would meet each other and the mating then would be successful because of the (C+G)% compatibility. Thus, in a freely interbreeding population (i.e., a species), there would be a tendency for small potential incipient species groups to form, from time to time, on the basis of random fluctuations (swings) in (C+G)% alone. This would provide one key component, the isolation component (sympatric), required for the emergence of a new species.

MUTATIONS IN "REPROTYPE" AND PHENOTYPE

Mutations changing (C+G)% (mutations affecting genotype) would not necessarily influence phenotype. In view of the role these mutations are here postulated to play cumulatively in reproductive isolation, they can be referred to as "reprotypic" mutations (i.e., mutations affecting the "reprotype"). Individual mutations affecting phenotype would only minimally affect (C+G)% (see example sequences in chapter 1), so that we will consider them as separate from reprotypic mutations. Speciation would most likely be successful when driven by both reprotypic and phenotypic mutations, only the phenotypic mutations, through changed gene products, being possible targets of conventional Darwinian natural selection (phenotypic selection).

Some significant phenotypic changes can be the result of a single mutation. A mutation (event 1), perhaps conferring greater speed and so allowing a horse more readily to escape from a predator, would only be likely to be accepted (a) if it happened to follow or accompany a (C+G)% fluctuation (event 2) conferring a degree of reproductive isolation for that horse from the major population, and (b) if the horse could encounter a mate with the similar (C+G)% fluctuation (event 3). The reproductive success guaranteed by the latter events, combined with the greater chances of survival based on the first event, would have created conditions most propitious for the origin of species.

Event 1 would often reflect more than a single mutation. For example, within a population of horses there is likely to be a bell-shaped distribution of speeds, just as there is likely to be a bell-shaped distribution of (C+G)%. Speed is likely to be a complex function requiring input from several genes. Certain compatible mutations in some of those genes would be shared by many of the faster horses. The concomitant events required to create the best conditions for speciation would be for the set of mutations corresponding to the small group of fast horses (at the margins of the bell-curve for speed), to also correspond within an individual with the set of mutations at the margins of the bell curve for (C+G)%. The chance of that individual finding a suitable mate, thus favouring perpetuation of the trait, might be greater if the phenotypic change depended on more than one mutation.

Thus, successful speciation might require mutations in each class. A (C+G)% swing (fluctuation) away from the species norm will require many mutations. Often phenotype swings, involving characters dependent on multiple genes, require several mutations. The requirement for more than one event might mean that speciation would sometimes appear gradually progressive, and not always, as Bateson would say, "discontinuous" (see chapter 13).

THE MULE

An example of two modern species in which the speciation process appears incomplete, may serve to illustrate some of the issues. Asses and horses are likely to share a common ancestor. The ass is more suited to some ecological niche than the horse, and vice versa. If a modern male ass mates with a female horse (mare), the product, a mule, is healthy but sterile. There is post-zygotic isolation between asses and horses due to mechanisms acting at the level of meiosis, which result in degenerate gonads in their offspring. Before the evolution of this post-zygotic isolation was complete prototypic mules would have shown some degree of fertility. The evolutionary fitness of prototypic horses would have been threatened not only by prototypic asses but also by the prototypic mules which succeeded in mating with a horse.

Today, despite the completion of post-zygotic isolation, asses and horses are still able to impair each others evolutionary fitness. One way this could occur would be through the loss of time available for reproduction with a more suitable partner. Given that the gestation period of a mare is about a year, and that the maximum number of pregnancies in a lifetime is unlikely to exceed 20, fertilization by one ass would result in a fitness loss of at least 19/20. When considering the evolution of horses one usually thinks of a selection pressure for speed in response to predators such as lions. It is not usual to think of asses in the same context as lions. Yet, given the gestation period of a mare and the possible number of pregnancies in a lifetime, fertilization by one ass could exert a considerable selection pressure. A mare which developed, say, an aversion to the smell of male asses (a pre-zygotic isolation factor), would enjoy a reproductive advantage of 20/20 versus the 19/20 of a mare without such an aversion.

Thus, having *previously* achieved post-zygotic isolation (the initial isolating mechanism), horses and asses (in the absence of interference by the breeder), would now be in the process of developing pre-zygotic isolation. Once such pre-zygotic isolation was achieved, then the initial selective pressure which had resulted in post-zygotic isolation, would diminish. If this involved divergent (C+G)%, then the values of the two species might then converge so that traces of the initial pressure would become less evident. The initial mechanism by which reproductive isolation had been achieved would be disguised. Since the initial isolating mechanism may diminish, it is appropriate to refer to later developing isolating mechanisms as "substituting," rather than "reinforcing" the initial isolating mechanism. These "reinforcing" mechanisms greatly concerned Wallace in his nineteen-point "proof" concerning the role of natural selection in hybrid sterility (see chapter 20).

MICROMUTATIONS AND MACROMUTATIONS

The usual initial isolation mechanism, differences in (C+G)%, can be considered to reflect changes at the level of individual bases (slowly accumulating micromutations), which are not obvious when chromosomes are examined microscopically. Later-developing "substituting" mechanisms would include suddenly-appearing macromutations (inversions, deletions, changes in chromosome numbers). Such features are observed in chromosome studies of modern hybrids (Chandley et al. 1974), and it is understandable that these obvious secondary chromosomal rearrangements, probably occurring *after* the initial evolutionary divergence was complete, should have received attention as candidates contributing to the initial isolation process (Bush et al. 1977; King 1993). Thus, White (1978 ch.10, 324) concluded: "Over 90% (and probably over 98%) of all speciation events are accompanied by karyotic changes [chromosomal macromutations], and ... in the majority of cases the structural chromosomal rearrangements have played a primary role in initiating divergence."

This requirement for macromutations has led some to dismiss speciation by chromosomal mechanisms in favour of alternative genic mechanisms (see chapter 7; Coyne and Orr 1998). The strength of the micromutation proposal is that it allows degrees of compatibility, so that a rare reproductive variant would be more likely to find a partially matching physiological complement. This would be unlikely if chromosomal speciation results from "single-step" macromutations (King 1993 ch.8, 92-104).

GENIC DIFFERENCES PRODUCE HYBRID INVIABILITY

To deal with various problems (e.g., "adaptive valleys"), the general form of the genic hypothesis begins with an organism homozygous for genes

A and *B* (Dobzhansky 1937 ch.8, 256). The alleles, one on the paternally-derived chromosome and one at the same position on the corresponding maternally-derived chromosome, may be represented as *AA* and *BB*. The genotype is written as *AABB*. The products of the genes do not impede each others functions. Let an *A* allele undergo a base change to a form we may represent as *A'*. Similarly, a *B* allele changes to *B'*. If it so happens that the products of *A'* and *B'* interfere with each other's function, then the heterozygotes *AABB'* and *AA'BB* will be healthy, but heterozygotes *AA'BB'* will malfunction. The incompatibility between *A'* and *B'* will favour divergence into two populations.

This genic difference will not necessarily involve reproductive functions such as those discussed by Bateson (anther mutants; see chapter 4) and Romanes (flowering-time mutants; see chapter 5). Since most genes are *not* primarily involved in reproduction, the general form of the genic hypothesis refers mainly to non-reproductive functions. Thus, the most likely result of genic incompatibility would be hybrid inviability. Should hybrid sterility occur, it would not be "cured" by allotetraploidy (see chapter 7).

AN IDEAL EXPERIMENT

An ideal experiment to obtain evidence for the proposed role of (C+G)% in speciation, would be progressively to change the (C+G)% of an organism *A*, to generate organisms *A'*, *A''*, *A'''*, etc. The abilities of these lines to produce fertile offspring with *A* would then be studied. It would be shown that above a critical level of (C+G)% difference the offspring would be sterile. The changes would not be specific to a particular gene locus, and could be contributed by disparate parts of a chromosome.

Actually, something remarkably like this experiment has already been carried out. Using the preferential sterility of one sex as an index of incipient speciation ("Haldane's rule"; see chapter 19), Naveira and Maside (1998) introduced genomic segments from one fruitfly species (*B*) into another fruitfly species (*A*), to generate variants *A'*, *A''*, etc.. The distinct "species" *A* and *B* were closely related, but crosses produced infertile males, and the incidence of these could be used as an index of incipient speciation. When crosses were made between *A* and *A'*, between *A* and *A''*, etc., it was found that infertility was obtained when about 40% of an *A* chromosome was replaced with part of the corresponding *B* chromosome. Remarkably, the 40% did not have to consist of one chromosome segment, but could be constituted from any of a number of subsegments.

The authors referred to their results as "unexpected," and suggested that "a new paradigm is emerging, which will force us ... to revise many conclusions of past studies." However, their interpretation, although hostile to the purely "genic" viewpoint, was cautious:

Current evidence suggests that the genetic basis of hybrid male sterility in *Drosophila* is generally polygenic, and the total number of sterility factors must probably be numbered at least in the hundreds. The individual effect on fertility of any of these factors is virtually undetectable, but it can be accumulated to others. So hybrid male sterility results from the ... [introduction] of a minimum number of randomly dispersed factors (polygenic combination). The different factors linked to the X [chromosome] on the one hand, and to the autosomes on the other, are interchangeable.

They also introduced non-coding DNA, concluding: "The effect detected after inserting non-coding DNA suggests that the coding potential of the ... [DNA] might be irrelevant for hybrid male fertility. It might be only a question of foreign DNA amount."

So what makes a DNA, as DNA per se, "foreign"? It might be the presence or absence of secondary modification (e.g., methylation). It might be a characteristic frequency of a particular subsequence (Bell et al. 1998; Lao and Forsdyke 2000b). It might be the (C+G)%. A "polygenic" explanation implies interaction between multiple gene products, but the data appear to support "chromosomal" explanations quite well.

SUMMARY

The major role of sexual reproduction is to allow meiosis to occur. Here, father's and mother's contributions to your genome are compared. There is then the opportunity for recombination repair so that fewer deleterious mutations will be conveyed by your gametes to the next generation. The species-dependent component of the base composition provides an index of whether your father's genome is likely to be a good template for repair of your mother's, and vice versa. A divergence between your parents in this component might make you sterile (a mule). Assuming (say) that your father's genome had diverged substantially from the population norm, it is possible that within the population there will be potential female partners with the *same* divergence, with whom he could have reproduced to give you non-sterile step-siblings. This is the basic precondition for the initiation of speciation. Initial reproductive isolation would reflect slowly accumulating "micromutations" which change the species-specific component of base composition, (C+G)%. Later developing substituting mechanisms would include suddenly appearing "macromutations" (e.g., inversions, deletions).

13 Relationship to Physiological Selection

"Although the leading English biologists are not hopeful material to convert, no doubt the rising generation will prove better able to distinguish the fact that two and two equals four ... ."
Letter from Romanes to Gulick, 1889 (Gulick 1932 ch.15, 417).

Chapters 9 to 12 update Darwin's *The Origin of Species* in the light of recent advances in molecular biology and bioinformatics. It remains to consider how the new viewpoint relates to the process of physiological selection as proposed by Romanes (Part 1). It will be for the reader (and posterity) to decide whether two and two equals four. Bateson's best attempt to explain the facts of hybrid sterility (chapter 6) provides a convenient point of departure.

BATESON REVISITED

"If species have a common origin, where did they pick up the ingredients which produce this sexual incompatibility? Almost certainly it is a variation in which something has been *added*" (Bateson 1922a). The added something would be the deviation in (C+G)%, slowly "proceeding by the accumulation of indefinite and insensible steps [point mutations]" to produce the *genome phenotype* (i.e., something detected only when genome meets genome, namely hybrid sterility). The "complementary factors" of Bateson "which need not, and probably would not, produce any other perceptible effects" (given the degeneracy of the genetic code) would be the (C+G) percentages of the paternal and maternal genomes. The "substance" formed in the hybrid by the meeting of two such "complementary factors" would be a recombination complex between the two genomes so fragile, because of (C+G)% incompatibilities, that it would "abort" and hybrid sterility (due to defective meiosis) would result. In Bateson's words (1909a, 230):

Failure to divide is, we may feel sure, the immediate "cause" of ... [hybrid] sterility. Now, although we know very little about the heredity of ... differences, all that we do know points to the conclusion that the less-divided is dominant to the more-divided, and we are thus justified as supposing that there are factors which can arrest or prevent cell division [see Murray 1992; Page and Orr-Weaver 1996]. My conjecture therefore is that in the case of sterility of cross-breds we see the effect produced by a complementary pair of such factors [Bateson's italics].

Thus, the (C+G) percentages of the parental genomes, having drifted apart by random mutation, would not be *sufficiently* similar to allow efficient templating for recombination repair. Meiosis would not be allowed to proceed and the hybrid sterility genome phenotype would appear. When accompanied by mutations affecting the conventional phenotype, conditions for classical phenotypic ("natural") selection would have been created, as set out in the previous chapter.

For Bateson (1913 ch.11, 242) the agenda was quite clear: "The first step is to discover the nature of the factors which by their complementary action inhibit the critical divisions and so cause sterility of the hybrid." Thus, Bateson was calling for the discovery of what we now know as DNA. Bateson's careful use of the general term "factors" (not "genes") allowed him to encompass any form of information (primary or secondary) which was contained in the chromosomes or elsewhere (chapter 6).

He tended to refer to genes as responsible for "transferable characters," which we here interpret as primary information. He did not consider that "any accumulation of characters" of this order "would cumulate in the production of distinct species." Yet these transferable characters Bateson (1922a) regarded as "attached" to the same "base" as factors responsible for "specific difference" (here interpreted as secondary information). The distinctness of the latter factors was explicit. Under the heading "Possible limits to recombination" Bateson (1909b ch.3, 73) noted:

It will probably occur to many that there are limits to these possibilities of transference [of genic characters], and so undoubtedly there are. The detection of these limits [which include barriers affecting speciation] is one of the more important tasks still awaiting us. ... We must surely expect that these transferable characters [conventional genic characters] are attached to, implanted upon, some basal organization, and the attributes or powers which collectively form that residue may perhaps be distinguishable from the transferable qualities. *The detection of the limits thus set* [by the basal organization or residue] *upon the inter-changeability of characters* [from parents to offspring over the generations] *would be a discovery of high importance and would have a most direct bearing on the problem of the ultimate nature of Species.*

Only two years before his death he reiterated (Bateson 1924): "We have determined the transferable characters as one group, and we no longer confound them [confuse them] with the essential elements conferring specificity [speciation]. Segregation [of characters] is of course often seen in species crosses, but as to the behaviour of these critical elements [those conferring species specificity] we know as yet very little."

Given the state of knowledge at that time, it is hard to imagine how Bateson could have expressed this more precisely. His "critical elements" would correspond with the "intrinsic" (non-genic) gonadal "peculiarity" responsible for the reproductive isolation proposed by Romanes (1886; see chapter 5). If we had to fault Bateson, we might observe that he was unwilling to affirm that the critical elements might segregate at meiosis (he invoked "non-Mendelian phenomena"), and that, although he assigned genes and the critical elements to the *same* "base," and was prepared to consider factors responsible for phenotypic characters (genes) as associated with chromosomes, he was not prepared to assign his critical elements to chromosomes. Perhaps this was a tactic to prevent any confusion between genes and the critical elements.

CREATION BY VARIATION MAY APPEAR DISCONTINUOUS

Bateson (1909a, 230) had affirmed: "that the 'Origin of Variation,' whatever it is, is the only true 'Origin of Species,'" and he thought that "definite variation" would tend to arise in "discontinuous steps." This might occur in the case of species characterized by traits involving relatively few genes (e.g., tallness and dwarfness in peas), but, as indicated in the previous chapter, the speciation process could potentially occur more smoothly in the case of multigenic traits. Thus Bateson and Saunders (1902) noted:

It must be recognized that in, for example, the stature of a civilized race of man, a typically continuous character, there must certainly be on any hypothesis more than one pair of possible allelomorphs [alleles]. There may be many such pairs [within an individual], but we have no certainty that the number of such pairs, and consequently of the different kinds of gametes, are altogether *unlimited* even in regard to stature. If there were even so few as, say, four or five pairs of possible allelomorphs, the various homo- and hetero-zygous combinations might, on seriation, give so near an approach to a continuous curve, that the purity of the elements would be unsuspected [Bateson's italics].

A swing (be it a swing in the genomic phenotype (C+G)%, or a swing in some phenotypic character which results from the interaction of numerous genes), is the cumulative result of the gradual accumulation of many mutations. For the speciation process to begin, two swings (the extremes of the corresponding bell curves) would have to correspond. If this were a relatively rare event, then evolutionary advance might sometimes appear irregular, or punctuated. Once a form of isolation is available, phenotypic change can occur very rapidly, so that any intermediate forms might persist for only few generations.

If, in the evolutionary process, the initial creation of conditions for phenotypic variants to arise (i.e., the production of small isolates of distinct percentage C+G), rather than their subsequent natural selection, were a major limiting factor, then as noted in chapter 3, large populations occupying large areas would be more likely to produce such variants. Darwin wrote (1859 ch.4, 105): "I am inclined to believe that largeness of area is of more importance, more especially in the production of species, which will prove capable of enduring for a long period, and of spreading widely. Throughout a great and open area ... there [will] be a better chance of favourable variations arising from the large number of individuals of the same species there supported."

However, Darwin maintained (1859 ch.4, 106) that *natural selection*, rather than the *production of variants*, is rate-limiting:

> Finally, I conclude that, although small isolated areas probably have been in some respects highly favourable for the production of new species, yet ... the course of modification [by natural selection] will generally have been more rapid in large areas; and what is more important, that the new forms provided in large areas, which have already been victorious over many competitors, will be those that will spread more widely, [and] will give rise to most new varieties and species.

He often turned to "the smaller continent of Australia" for examples. Yet, the land mass of Australia is very large, and, at least in modern times, covers a wide range of habitats. Lyell (1863 ch.21, 414) noted: "There is generally an intimate connection between the living animals and plants of each great division of the globe and the extinct fauna and flora of the post-tertiary or tertiary formations of the same region. ... We find in Australia, not only living kangaroos and wombats, but fossil individuals of extinct species of the same genera."

No members of the cat, rabbit, or prickly pear families, appeared in this large area. We cannot say definitively that the limiting factor was not selection, but certainly we do know that, following the introduction of the rabbit, cat and prickly pear into Australia in recent centuries, natural selection offered little impediment to their proliferation. It seems most likely that it was the initial "discontinuous" emergence of the required variant groups (depending on Romanes' "three pillars" of variation, heredity, and reproductive isolation) which was rate-limiting. Natural selection of domestic variants, as breeders have known for centuries, is quite rapid *provided* there is reproductive isolation (in this case provided by the breeder). Without that isolation, natural selection may be frustrated.

DIFFERENCE BETWEEN VARIETIES AND TRUE SPECIES

Huxley and Bateson were much concerned that generation of varieties under domestication was not an adequate model for the generation of species under natural conditions (see chapter 3). If the breeder assumes the role of pre-zygotic isolator, then clearly there is no selective pressure for post-zygotic isolation involving the generation of differences in (C+G)%. Thus, hybrids between different varieties, which may differ quite dramatically in the morphological characteristics selected by the breeder, would be expected to be perfectly fertile. Darwin, of course, would not have known about (C+G)%, but he does not appear to have been too far from the mark in stating that hybrid sterility: "has arisen incidentally during ... [the] slow formation [of species] in connection with other and *unknown changes in their organization*" (see chapter 3). In agreement with Hooker, Mendel, Nägeli, and others, Bateson (1913 ch.1, 30) also may have been prescient in noting that "the phenomenon of variation and stability must be an index of the *internal constitution of organisms*, and not mere consequences of their relations to the outer world."

INBREEDING AND OUTBREEDING

Your father and mother united to generate you (their hybrid) so that on some future occasion within your gonad the near-identical information in their genomes could be compared at meiosis. This would allow error-correction, so that you would be able to produce near-perfect gametes. To ensure that their genomes are suitable for this mutual error-correcting task, the genomes are first checked for "accent" compatibility [(C+G)%]. Small differences in this characteristic would disrupt meiotic pairing, and hybrid sterility (infertility) results.

The lack of vigour resulting from inbreeding (Figure 8.1), is most readily explained if closely-related individuals, sharing many mutations at *identical* positions in their DNAs, were unable to offer a sufficiently diverse template for error-correction during meiosis, or for elimination of mutations by recombination and subsequent purifying selection (Winge 1917). The resulting gametes would tend to perpetuate the defects of their parents. The converse of this is the hybrid vigour resulting from within-species outbreeding. It should be noted that we are primarily concerned with the germ-line rather than with the soma. The possibility that genomes may cooperate to bring about repair in somatic cells is not considered here (Bernstein and Bernstein 1991 ch.12, 255).

NEUTRALISM

Since mutations can affect both the conventional phenotype and the genome phenotype, one must be careful in describing a mutation as "neutral" (chapter 11). When sheltered by some form of reproductive isolation apparently non-adaptive mutants can emerge as new species. Thus Gulick noted snails with minor differences in shell markings (Gulick 1872a,b; see chapter 5). In this case, since the isolation was geographical, it was possible that there was no accompanying physiological (genomic) isolation, so that snails from different valleys might still be successfully crossed. However, when isolation is genomic there remains the possibility that the mutation causing the change in shell marking (non-adaptive phenotypic effect) was also making some contribution (albeit small) to genomic isolation (variation in percentage C+G). Thus, although phenotypically neutral with respect to natural selection, the mutation might not have been genotypically (reprotypically) neutral. In Part 3 we consider other genetic phenomena (e.g., "purine loading"; see chapter 15), which cast further doubt on the idea that any accepted mutation is neutral (Forsdyke and Mortimer 2000).

SUMMARY OF THE ORIGIN OF SPECIES

There is one *general* mechanism of speciation which applies to most taxonomic groups. DNA carries *primary* information (its primary sequence) which encodes characteristics (the conventional phenotype) that may be the target of phenotypic selection (conventional Darwinian "natural" selection). DNA also carries *secondary* information, the genome phenotype, which may be the target of the most evolutionarily-significant form of reproductive selection (Romanesian "physiological" selection). This is manifest when genome meets genome in the process of recombination, and causes hybrid sterility in the case of inter-species crosses.

The relevant secondary information is the species-dependent component of the base composition [(C+G)%], which tends to be a dispersed genome-wide characteristic. (C+G)%, the "accent" of DNA, affects stem-loop potential, thus playing a critical role in the initial (paranemic) homology search preceding recombination. The product of the quantities of C and G is the major determinant of stem-loop potential, so that small fluctuations in C and G should have a major impact on the success of a homology search. Thus, compatibility in (C+G) percentages is required for recombination between DNA molecules.

As the result of random fluctuations in (C+G)% within a species, various interbreeding sub-groups (potential incipient species) will emerge from time-to-time in the same geographical area (sympatrically). Depending on the size of the sub-groups, operation of classical Delboeufian random drift alone will ensure that subgroups will differ in characters from the parental group. These characters may be non-adaptive (phenotypically "neutral") or adaptive. If adaptive, they will be subject to phenotypic selection, and this will greatly accelerate the evolution of the adaptive characters. The species which emerges will, by definition, be reproductively isolated. Further differentiation into higher taxonomic groupings will occur through further speciation events and conventional Darwinian natural selection.

Thus, it seems probable that (C+G) percentages of meiotically pairing chromosomes are the modern chemical equivalent of the "other and unknown changes" of Darwin, of the peculiar "physiological complements" of Romanes, and of the "complementary factors" of Bateson.

SUMMARY

A revisitation of the analyses of Darwin, Romanes, and Bateson indicates that each, in his time, went as far as was possible in stating the truth about the origin of species, incomprehensible as it may have seemed to their contemporaries (and to some modern evolutionists). Romanes' "physiological complements" and Bateson's "complementary factors" are two sides of the same coin, namely the species-dependent component of DNA base composition. The Physiological Selection Theory is consistent with Bateson's insistence on the discontinuity of observed variations, and with a punctuated pattern of evolutionary progression.

PART THREE

The Divergence and Convergence of Species

14 Selfish Genes and Selfish Groups

"It is a probable hypothesis, that what the world is to organisms in general, each organism is to the molecules of which it is composed. Multitudes of these, having diverse tendencies, are competing with each other for opportunities to exist and multiply; and the organism, as a whole, is as much a product of the molecules which are victorious as the Fauna or Flora of a country is the product of the victorious beings in it. On this hypothesis, heredity transmission is the result of the victory of particular molecules contained in the impregnated germ. Adaptation to conditions is the result of the favouring of the multiplication of those molecules whose organizing tendencies are most in harmony with such conditions."

T. H. Huxley (1869b ch.4, 115)

We were disconcerted when Copernicus told us, what the ancient Greeks had long thought, that the earth is not the centre of the universe. A major lesson of the ongoing revolution in the biological sciences is even more disconcerting. Phrased in Copernican terms we are told that we are not at the "centre" of our own DNA. The DNA in each of our cells encodes the information for many of our characteristics, such as the colour of our eyes, or whether we are likely to become diabetic. Our DNA occupies the central and most hallowed part of our cells, the nucleus. Here it is guarded by membranes and vigilant watchdog enzymes that protect and repair. Yet, we are not at the "centre" of our own DNA?

To astronomers the earth does not appear central in the sense that, by most frames of reference, the sun and planets do not revolve around the earth. We are on a small planet, in a small solar system, in a minor galaxy, which is one among many millions of such galaxies. In the face of the enormity of it all, the earth appears quite insignificant. The earth does not appear of major concern to the rest of the universe, and *we* do not appear of major concern to our DNA.

How can that be? Our DNA is within our bodies, not out there in space. Each of our perhaps a hundred thousand billion cells contains an approximately identical DNA "book" of instructions. The idea that

"chapters" of the book might be read selectively to provide the information necessary for different types of body function can be traced back to
the nineteenth century (De Vries 1889; see chapter 15). A liver cell is a liver
cell because in that cell the "liver chapter" of DNA is selectively read. If
we, by accident or design, succeed in destroying the earth in a thermonuclear explosion, the universe will not even blink. If you, by accident
or design, destroy yourself, all the cells of your body will die and with
them the DNA in the nuclei. How can we not be central if the fate of the
DNA we contain hangs so perilously at risk? The answer is that if you
destroy yourself, DNA will not even blink.

WHOSE AGENDA?

If we destroy the earth in a thermonuclear explosion, your DNA will be
lost for ever. Until that time comes, your DNA ensures its survival by
"distributing" (encoding functions which facilitate distribution) copies
of itself among other members of your species on the planet. It is true
that the particular DNA within you is unique in the sense that the "deck"
of gene cards has been shuffled in a particular way to produce you. You
should rightly rejoice in your individuality. But if you were to die we
could reconstruct many of the features of your "deck" by examining the
"decks" of your near relatives. Of course we could not easily reconstruct
you, because you are the product both of your genes and of your environment. The essence of *you* is information. You feel unique because of
the particular set of memories gathered in a unique order at particular
times and places. But DNA is not greatly concerned with details imposed
by environment during our somatic existence.

Our insignificance with respect to DNA seemed to be evident when it
was found that the "chapters" in our DNA which appear to affect us occupy no more space than does a brief appendix to a book. Only about 2% of
our DNA encodes information of immediate relevance to our existence.
The rest of the DNA appears just "along for the ride." Some have disdain-
fully labelled it "junk DNA." But who is to say that the anthropocentric
view is correct? The earth is no longer perceived as the centre of the universe, but it does not automatically follow that we should label Mars,
Pluto, Venus, and the rest, as "junk." Indeed, it is possible that junk DNA
has survived for the purpose of intracellular defence (see chapter 17).

The book you are now reading implies a purpose which is easily recognized. The author had something he wanted to say and chose the
medium of the book, rather than soap-box oratory, or appearance on TV.
The title and every chapter subserve this purpose. But there are other
purposes. There are other agendas. A book is a paper manufacturer's way
of selling paper. The paper manufacturer usually cares little about the

subject of the book so long as there are people out there willing to buy paper. The addition to the paper of varied combinations of 26 inked symbols helps achieve the paper manufacturer's agenda. The reader usually is not concerned with a few blank pages at the beginning and end of the text; but if 98% were blank and only 2% were text there would be concern. It is argued that the book of DNA has its own agenda. Unless we can recognize a function for the 98%, we must concede that our concerns may not be its concerns.

When considering the matter it is reasonable to ask whether, from the viewpoint of evolution, one agenda is more important than another. When trying for a given biological situation to determine which agenda is being best served, it is appropriate to apply what the economists refer to as "marginal utility," and which we might call the "crunch test." Would the paper production system break down if the paper manufacturer were denied the privilege of publishing books with a few blank pages at the beginning and end of the text? Probably not. The paper manufacturer could absorb this small decline in paper sales. On the other hand a paper manufacture who tried to foist 98% of blank pages would not fare well. When the crunch came, the paper manufacturer's agenda would have to concede to that of the customer.

WINNING AND LOSING

Darwin (1875 ch.19, 166) noted *inter*species competition between gametes (pollen): "If pollen of a distinct [another] species be placed on the stigma of a flower, and its own pollen [from a member of its own species] be afterwards, even after a considerable interval of time, placed on the same stigma, its [own pollen's] action is so strongly prepotent [dominant] that it generally annihilates the effect of the foreign pollen."

Mendel (Iltis 1932 ch.14, 202) noted *intra*species competition: "Pollination with a number of pollen grains is advantageous because there is rivalry between pollen grains when the stigma is dusted with a plurality of these, a rivalry in which the most vigorous conquers, whereas when the stigma is dusted with only one pollen grain this may result in the production of nothing but a weakly seed, or even no seed at all."

An implication of this was pointed out by Haldane (1932 ch.5, 123): "A higher plant species is at the mercy of its pollen grains. A [mutant] gene which greatly accelerates pollen tube growth will spread through a species [enhance it's own survival] even if it causes moderately disadvantageous changes in the adult plant. [Conversely] a gene producing changes which would be valuable in the adult would be unable to spread through a community if it slows down pollen tube growth." In this case the agenda of a gene dominates.

As mentioned in chapter 3, Darwin raised the issue of how a process (hybrid sterility), which appeared disadvantageous to an individual (encoded by a group of genes), yet benefitted the species (a group of individuals), could have evolved. It appeared that in this case the agenda of the group of individuals dominated ("group selection"; see chapter 5). In both cases, individuals seem to lose out, either to the collective of individuals (the species), or to certain members of the collective of genes within individuals. By "winning," we generally mean survival. Species and genes tend to endure, individuals to die. One might argue that species and genes do not seem to have much fun, and so can hardly be considered to have won. Perhaps fun is the price of our mortality?

PANGENS

In 1893 (preface, vi) Huxley summed up his views on evolution: "We still remain very much in the dark about the causes of variation; the apparent inheritance of acquired characters in some cases; and *the struggle for existence within the organism*, which probably lies at the bottom of both of these phenomena." Indeed, Darwin (1875 ch.27, 399) had noted: "Each animal and plant may be compared with a bed of soil full of seeds, some of which soon germinate, some lie dormant for a period, whilst others perish. ... An organic being is a microcosm – a little universe, formed of a host of self-propagating organisms, inconceivably minute and numerous as the stars in the heavens."

In developing his "provisional hypothesis of pangenesis," Darwin (1875 ch.27, 349) had been influenced by reports of the great proliferative powers of micro-organisms (see chapter 9). The hypothesis required modification by Hugo de Vries in his *Intracellular Pangenesis* (1889) to make it approximate to modern concepts of the gene (see chapter 15). However, Darwin's idea that "gemmules" could transfer from one cell to another, where they could become part of the genetic material of the new cell (Figure 4.2), finds a modern analogy in the phenomenon of viral latency, where viral nucleic acid "seamlessly" integrates and hides in the genome of its host. In the case of HIV, the virus is not known to enter the germ-line. Nevertheless, our genomes are littered with retroviral remnants, indicating that, in the past, HIV-like viruses, similar to Darwin's proposed "gemmules," have indeed transferred somatically-acquired information to the germ cells, and hence to the offspring.

ALTRUISM

In *The Selfish Gene,* Dawkins (1976) described a world of DNA segments (replicators or genes), each "competing" in Darwinian fashion to get aboard the winning gamete for the trip through the "bottleneck" into the next generation. To better achieve this, groups of replicators have "built" themselves "survival machines" (cells and bodies). Of course, genes do not themselves directly "compete" or "build," but this avoids long-winded phrases such as: "segments of DNA encoding (say) cell wall components, will tend to survive and reproduce more, when compared with naked DNA segments which do not encode such components."

This "gene's eye view" of biological evolution provided an explanation for the evolution of altruism (actions which benefit others at the expense of the individual altruist) in terms of "kin selection." Since your kin are likely to have similar genes as yourself, a gene which leads you to sacrifice yourself, but facilitates the survival of three siblings (each likely to have at least 50% of your genes), will cause more copies of itself (the gene) to enter the next generation. Thus, a gene encoding products which cause you (consciously, or unconsciously) to act altruistically may perpetuate and increase itself.

Although this appears a simple example of the dominance of the gene agenda, group selection plays a role. Usually, your kin have to be *near* you to benefit from your altruistic conduct. So some degree of geographical isolation of your kin, relative to other members of your species, must *precede* your altruistic conduct. The conduct will confer an advantage on your kin group, over other groups which may not have the genes for altruism. In the "struggle for existence" between groups, your group will prosper (Gould 1998). Here we see *both* gene agendas ("selfish genes") and group agendas ("selfish groups") dominating individual agendas.

With regard to altruism, Haldane proposed (1932 ch.5, 131): "For in so far as it makes for survival of one's descendants and near relations, altruistic behaviour is a kind of Darwinian fitness, and may be expected to spread as the result of natural selection." However, we should note that the group isolation which must occur *first* might *not* involve natural (phenotypic) selection. The advantages of hybrid vigour (chapter 8) might favour genes which encourage the geographical dispersal and mingling of members of a species. Counteracting genes might appear from time to time so that related organisms would flock together, or at least not migrate away from each other. This random event (event 1) would create conditions such that genes for altruistic conduct could prosper under the influence of natural selection (event 2). In the absence of the coincident event 2, genes for aggregation into flocks with near-relatives might decrease fitness.

We should also consider here Batesonian "factors" (chapters 6 and 13), instead of genes encoding specific functions. At the *initiation* of the speciation process an organism's genome-wide (C+G)% broadly *defines* its kin, since this is what will ensure reproductive isolation (chapters 11 and

12). $(C+G)\%$ is a randomly-emerging "factor" necessary for group (kin) selection. The tendency of $(C+G)\%$ to self-perpetuate is greatly increased if genes encoding products which favour survival and reproduction (so being the target of conventional natural selection), become associated with it (Forsdyke and Mortimer 2000).

Initially the group selected for preferential survival because of the altruistic conduct of its members may be small. In this case the pressure for survival of genes for altruism will be high, because the degree of kin relationship will be high. As the selected group gets larger, the probability of genetic relatedness with other members of the group will decrease, so the benefits of altruistic conduct will be lessened. However, the genes for altruism may by then have become established. Unless actively selected against, they will persist just as vestigial organs persist (Romanes 1874).

MICRO-ORGANISMS FACE THE CRUNCH

So, you are your DNA's survival machine. Your DNA replicators have "created" you because you will facilitate their perpetuation. Maybe much of *you* could be created at little cost to the replicators (i.e., perhaps only 2% of genome space was needed). But it has not always been that easy. In the course of evolution certain micro-organisms got themselves into situations where speed of genome replication, or being compact in order to facilitate transfer, became of critical importance. It is difficult to replicate rapidly, or to be compact, and at the same time maintain a large quantity of DNA which does not serve these ends. The conflict seems to have been resolved by dumping the "junk." When the evolutionary crunch came, the load was lightened. When we look at the DNAs of viruses, bacteria and yeast, we find that most of it is dedicated to obvious function. There is little, if any, DNA for which an overt function cannot be assigned. For a replicator to survive in micro-organisms it also has to contribute to the immediate function of its survival machine.

Despite this streamlining of the genome, for some impatient DNA replicators the gentle pace of collective movement through the generations with other DNA segments ("vertical transmission") was not enough. They have broken away from the survival machines in which they initially evolved and have "built" themselves coats to protect themselves temporarily while transferring to a new host ("horizontal transmission"). DNA segments not critical to their survival were dumped by these newly diverging species, which we call viruses. Their evolution has now become critically dependent on successful transfer from host to host.

As the result of divergence among viral species, new virus species were created. In some cases viruses became integrated into the DNA of the somatic cells of their hosts, replicating their DNA as part of host cell

DNA replication. Here they came to lie in a "dormant" or "latent" state. But the host would eventually die, so their period of tenure was finite. At some point they would have to move on and seek a new host.

Some viruses actually penetrated the germ-line. Insertion into germ-line DNA was often undetected. Survival was assured as long as the host itself was reproductively viable. Thus replicators have both diverged to generate new viruses, and converged when extrinsic viruses become part of the germ-line of a new host. If the newly acquired DNA is not manifestly harmful it may be accepted, just as "junk" DNA appears to be accepted. It may preferentially insert itself into a host DNA region with a similar (C+G)%, or, once inserted, it may "camouflage" itself by adopting a (C+G)% characteristic of the region (Rynditch et al. 1998).

EMERGENT PROPERTIES

Dorothy and her companions were surprised and dismayed when at the end of their long search the Wizard of Oz turned out to be a humbug (Baum 1990). The same feeling may arise when one realizes that at the heart of biology are just a few squabbling DNA molecules. Where is the "grandeur" which led Darwin (1859 ch.14, 490) to conclude that "from so simple a beginning endless forms most beautiful and most wonderful have been, and are being, evolved"? The concept of "emergent properties" may help.

Simple objects tend to have simple properties. A brick alone has a limited set of properties. Combine it with more bricks and new properties emerge. Combine it with some wood and the range of new properties becomes even greater. When objects increase in number and variety, emergent properties may appear as a result of interactions. These properties may not have been apparent in the original simple objects.

Life is deemed to have initially arisen in a "primeval soup" of molecules. At one unique time-point a molecular complex is deemed to have emerged with a *new* property, that of self-replication. The initial rate of replication may have been incredibly slow, perhaps once in a hundred years. But that was some four billion years ago, and there was no hurry. When it had replicated once there were two molecular complexes. Depending on their stability, a hundred years later there might have been four. Eventually there would have been many molecular complexes and some complexes might come to vary from others. Thus the Darwinian story began.

It is convenient to call these early molecules "replicators" and imagine them to be something like RNA (Joyce and Orgel 1993). As the result of inherent instabilities, or attack by physical or chemical agents, replicators would vary, often for the worse. However, a replicator which, as the result of some chance variation, could now replicate in ninety-nine years would seem to have won a selective advantage. The faster replication would be of no advantage if the new form were less stable, or if the process of replication were less accurate. Thus there would have been competition with respect to three properties – speed of replication, accuracy of replication, and stability. A replicator which by chance happened to optimize these would "win." This would be a process of linear (not branching) adaptation (Figure 5.1a). The successful replicator (asexual) would merely out-populate the competitors, perhaps by garnering more of the store of building-block molecules needed for its assembly. Unless a barrier to competition appeared (e.g., discovery of an appropriate ecological niche), no alternative form could survive. This is evolution in classical Darwinian mode. The number of replicator "species" would be no more than the number of ecological niches.

EXPLORING SEQUENCE SPACE

We will digress here to note that all organisms that exist, or have *ever* existed, are likely to represent but a minute fraction of potential "sequence space." For nucleic acids with four bases there are $4^1 (= 4)$ possible single base sequences, $4^2 (= 16)$ possible dinucleotide sequences, and $4^3 (= 64)$ possible trinucleotide sequences. Thus for the 4,600,000 nucleotide sequence of the bacterium *E. coli*, there are $4^{4,600,000}$ other combinations of the same bases. Some of these alternatives will have been explored and rejected. However, many alternatives, with the potential for more emergent properties, have never been "examined" by natural selection. In the past, exploration of sequence space has been limited by what happened to be randomly produced in the time available and by what was biologically feasible; but advances in biotechnology now permit the creation of sequences which have had no previous existence on earth.

A hint that potential for the bizarre may exist in sequence space is provided by "*pro*teinaceous *in*fectious particles" or "*prions*" (the vowels were transposed for euphony). This is the name given to a normal protein which has two possible conformations, one common and one very rare (Prusiner 1997). However, the rare form has the strange property of being able to cause the common form to change conformation to the rare form. Any property associated with the rare form is transferred between individuals by the rare form, with a pattern of inheritance similar to that normally requiring the transfer of conventional nucleic acid information. Solubility is an important protein property, and as the limit of solubility is approached, molecules of a protein will tend to aggregate (like-with-like). Whereas the rate of cellular production and decay of the normal resident protein ("self") is such that there is no aggregation problem, the conversion of the normal protein into the rare form by an invading copy of the rare form ("not-self"), greatly increases the concentration of the rare form, which, being less soluble, aggregates. Various transmissible neurological diseases, such as scrapie in sheep and mad cow disease, are caused by prions. All this is very unpleasant. However, a process by which pre-existing "self" molecules can assist the intracellular recognition of extrinsic "not-self" molecules (manifest as aggregation), might sometimes be turned to an organism's advantage, as will be considered in chapters 15 and 17.

RECOMBINING REPLICATORS NEED (C+G)%

At some stage in biological evolution, probably quite early, a highly beneficial property emerged. A replicator acquired the ability to exchange parts of itself with other replicators. Recombination (sex) had been "discovered." Now there was another route to a goal (structural integrity) which, up to that moment, had been served only by the properties of replication accuracy and chemical stability. A replicator which had partially degenerated as a result of replication error or random perturbation, could now be rescued or repaired. It could exchange segments with other replicators so that there was the possibility of reconstructing the original, non-degenerate, replicator complex. This may sound somewhat fanciful, but we know that RNA molecules *alone* can cut and splice themselves, so simulating aspects of this prototypic recombination process.

Thus a primeval soup of new "winners" would have emerged, the "recombining replicators." However, variation would have been ever-present, and this would have created the need for an important new property to emerge, self/not-self discrimination. A recombining replicator would have to *avoid* extreme variants and recombine only with its *own* type of recombining replicator in order to reconstruct its original form. Most extreme variants would not be superior to the original form. A rare variant which was superior in some way would have to recombine with a *similar* variant in order to sustain the superior property.

In this light it can be seen that, *if it were at all possible* (i.e., chemically and physiologically feasible) for recombination *and* self/not-self discrimination to have evolved collectively at an early stage, it is likely that these properties *would* have evolved. Could such apparently complex functions have evolved? The stem-loop "kissing" homology search between RNA molecules, as described in chapter 9, suggests how this primitive self/not-self discrimination might have arisen if the replicators were RNA-like. Thus, evolutionary pressures for the development of stem-loop potential (necessary for the homology search), and for differences in (C+G)% (necessary for self/not-self discrimination) might have arisen at an early stage. The smartest recombining replicators of the early "RNA world" would quickly have acquired stem-loops and appropriate levels of (C+G)%.

SURVIVAL MACHINES

Much of this could have occurred *before* the next major evolutionary step, namely the synthesis of proteins which would act as catalysts speeding up the processes required for generation of cell structure, for accurate replication, and for recombination. The genetic code relating each amino acid to a particular set of nucleic acid bases (codon) had to arise. How this came about is not clear (Szathmary 1999), but the result would have been that information for protein sequences would now have to *compete* for a place in nucleic acid sequences with information for stem-loops and (C+G)%.

At some stage the nucleic acid would have become double-stranded, and this would have greatly facilitated the monitoring of its integrity followed by repair as required. At some stage the nucleic acid would have changed from RNA to DNA, the former type of nucleic acid (as mRNA) being employed to transfer DNA information to the site of protein synthesis, and (as tRNA and rRNA) playing important accessory roles in protein synthesis. DNA not only had to encode stem-loops, the appropriate (C+G)%, and proteins, but also had to contain signals showing when to begin and end replication (DNA → DNA), and transcription (DNA → RNA). This involved regulation of access by proteins concerned with replication (e.g., DNA polymerase), and transcription (e.g., RNA polymerase). Thus, DNA has to convey through time and space *multiple* forms and levels of information.

OVERLAPPING LANGUAGES

Should you wish to mark a place in this book to facilitate subsequent access you might turn down a corner of the page. The flap is a message which you decode with the knowledge that it means "continue reading here." This is the information that the message conveys. You have imposed a new *form* of message on the page (i.e., the words "continue reading here" if written in the margin would contain the same information as the flap message). If the flap is small it may not conflict with the primary message (the print on the page). If it is large it may cover up part of the print. Thus the "secondary structure" of the page (encoding one form of information) interferes with your reading of the text (encoding another form of information). The print can still be read in its original form, but the flap has to be lifted to allow this.

In the case of the book there is a distinction between the medium (paper) and the message (contained in the print). You read (decode) the print to obtain information. In the case of DNA the "paper" is a phosphate-ribose chain, and the "print" is a sequence of bases. "Book-marks" in DNA can seldom avoid conflicting with one or more of the other messages the molecule is carrying. Some of these messages might influence the secondary structure of DNA. All this means possible conflict between several genome "languages." How many masters can DNA serve?

INTRONS

From this perspective it is possible to understand one of the most puzzling aspects of the coding of proteins by DNA, namely that the information for a protein is not continuous, but is interrupted by DNA sequences which do not contain protein information (introns). We are so familiar with the linear continuity of a written text, that it is difficult for us to imagine systems of information where this does not apply. When a file is saved in a computer, for example, the message may be scattered at a variety of codified locations throughout the disc. When you open the file the computer assembles the message in the linear form you require to obtain information from it. For all we know, your brain may then store the message in scattered codified locations, just as does a computer.

At the time of this writing (circa 1998) there is a lively debate on whether the interruptions in information for a protein were present when DNA molecules first acquired protein-encoding potential, or whether they were inserted into the protein-encoding sequence at a later date. The former scenario would be consistent with the idea of competition between various forms of information (messages) for space in the DNA sequence. Protein-encoding information might have had to *intrude* into sequences already adapted for genome-wide stem-loop potential and (C+G)%. The redundancy of the genetic code would have allowed such accommodation to (C+G)% (e.g., see the sample sequences in chapter 1). However, stem-loops, at least by virtue of the stems which require complementary base pairing (Figure 9.2) would place much tighter constraints. In this circumstance, it might have been evolutionarily advantageous to shift the stem-loops to non-coding regions, which we now call introns.

This hypothesis makes several predictions. (i) In view of the tight constraints on genome size in bacteria as discussed above, the intron option might not be available in these organisms. Indeed, bacteria have to accommodate both stem-loop potential and protein-encoding potential in the same sequence (Forsdyke 1995e); their protein-encoding sequences have no introns. (ii) Stem-loops should appear preferentially in introns, rather than in protein-encoding regions (exons). This is observed for some genes (Forsdyke 1995c). (iii) Genes whose protein-encoding capacity has to evolve very rapidly in response to intense phenotypic selective constraints, would have to abandon any attempt to co-encode stem-loops and protein in the same DNA region. Various genes under this type of positive Darwinian selection show this. Stem-loop potential is shifted to the introns (Forsdyke 1995d, 1996b). (iv) Rare genes which encode two proteins in overlapping regions (i.e., involving the same DNA sequence), would be less free to retain stem-loop potential in the region of the overlap. This is indeed found (Forsdyke 1995f; Barrette et al. 2001).

LOGICAL EXTENSION OF PRINCIPLE OF EMERGENCE

Haldane (1927, 286) considered that "the universe is not only queerer than we suppose, but queerer than we *can* suppose." Nevertheless, some would regard "mind" as the ultimate property to emerge on earth. Viewing emergence as a *process*, the anthropologist-philosopher-theologian Teilhard de Chardin (1959) extrapolated this in a controversial hypothesis of man's place in the universe, a topic which lies beyond the scope of this book. However, we should note that at each step in the process there is the opportunity for value judgements. Perhaps wood is best left as trees. Perhaps clay should be left in the earth and not fashioned into bricks. On the other hand, those who prefer not to live in caves would be pleased at the properties which emerge when wood and bricks are combined. The emergence of mind, in the sense conveyed by Huxley in the quotation at the beginning of the Prologue, suggests a human agenda-conflict between our biological purpose and a "higher" purpose (see chapter 5).

SUMMARY

This chapter summarizes, integrates, and extends the views of Hamilton, Williams, and Dawkins regarding conflicts between the apparent agendas of genes, individuals, and groups of individuals. Altruistic conduct, conventionally interpreted in terms of selfish genes, has a group selection component. While many organisms seem to accept a large part of their genome as "junk," bacteria and viruses have divested such DNA under pressures for compactness and speed of replication. If junk DNA has a function, then bacteria and viruses do not need this function, or can achieve it in other ways. Even so, only a small part of sequence space has been "examined" by natural selection. If they *could* have evolved (i.e., if chemically feasible), recombination and self/not-self discrimination *would* have evolved in the first emergent life forms ("replicators"). Adaptations supporting these processes would have preceded adaptations for the protein synthesis needed to construct "survival machines." Conflict between the several forms of information carried by DNA may explain introns.

15 Slow Fine-Tuning
of Sequences

> "Small steps in genotype space can have large consequences in
> phenotype space. ... As computer programmers well know, small
> changes in a complex system often lead to *far-reaching* and destruc-
> tive consequences (and computer programmers make these small
> changes *by design*, and with the hope of improving the code!).
>
> Lawrence Hunter (1993 ch.1, 3)

Most mutations are deleterious. Natural species appear so well adapted to
their contemporary environments that changes among species members
are generally for the worse. Changed members are selected against. So
well-adapted are many natural plant species that we designate them as
"weeds." Their members grow rapidly and form seed even in nutritionally unfavourable conditions. Domestic plants usually grow more slowly,
even in nutritionally favourable conditions.

Such observations suggest that the process which is the major concern
of this book, the *initiation* of speciation, is but the beginning of a much
slower process of adaptive fine-tuning. Following initial speciation
events, a multiplicity of secondary accommodations might then be necessary. If your violin is missing a string it is a major advance to acquire
and install a new string. A quick tune may result in an instrument capable
of a tolerable performance, but concert-grade work requires fine-tuning
which might be more protracted.

In biological evolution the asymptotic approach to perfect adaptation
might take many thousands of generations, even if the environment were
constant. This is something quite familiar to writers of computer software. A small change in one part of a program may have unforeseen consequences for other parts of the program. The laborious "debugging" of
programs largely involves the hunting down, correction, or accommodation to, such secondary effects.

Just as perfect adaptation may be approached very slowly, so may perfect
de-adaptation. Darwin was much puzzled by apparently non-functional
organs (e.g., nipples in males) which are designated as "vestigial" because

they resemble an organ which we see functional in some members of the same species (e.g., females), or in other species. The organ is often smaller than expected if it were functional. It appears to play no role in the economy of the organism. There seems to be no selection pressure for its retention, and it is on a path towards disappearance, but its progress along that path is very slow (Romanes 1874).

THE CROWDED CYTOSOL

Darwin's "Pangenesis" (Figure 4.2) made a lot of sense once cleansed by De Vries of the idea that gemmules might be transported to germ cells. Thus De Vries (1889, 215) noted:

> In the nucleus every kind of pangen [gene] of the given individual is represented; the remaining protoplasm in every cell contains chiefly *only* those that are to become active in it. ... With the exception of those kinds of pangens that become directly active in the nucleus, as for example those that dominate nuclear division, all the others have to leave the nucleus in order to become active. But most of the pangens of every sort remain in the nuclei, where they multiply [in modern parlance, replicate and transcribe], partly for the purpose of nuclear division, partly in order to pass on to the protoplasm. This delivery always involves *only* the kinds of pangens that have to begin to function [in the corresponding cell]. During this passage they can be transported by the currents of the protoplasm and carried into the various organs of the protoplasts. Here they unite with the pangens that are already present, multiply, and begin their activity. All the protoplasm consists of such pangens, derived at different times from the nucleus, together with their descendents.

We will now consider evolutionary fine-tuning at the level of intracellular molecules, with particular reference to the "pangen" RNA. Genes are transcribed to give a variety of RNA species, including a multiplicity of messenger RNAs (mRNAs; perhaps 10,000 distinct types per mammalian cell) and transfer RNAs (tRNAs; at least 20 types). Each tRNA corresponds to a particular amino acid, to which it can become transiently attached (Figure 15.1); bases at the tip of a loop in the tRNA form an anticodon sequence capable of pairing, rather like the inter-strand pairing in duplex DNA, with a particular three-base codon in mRNA. In this way the genetic code can be "read," and nucleic acid sequence information translated into protein sequence information. The tRNAs release their associated amino acids to form a chain of amino acids in the order specified by the sequence of codons in mRNA. This is the fundamental, unidirectional, link between genotype and phenotype. Some of the resulting proteins are exported from the cell, some form structural elements (e.g., cell wall), and some remain in the cell. Protein and RNA molecules are present in the cytosol (cytoplasm), where protein synthesis occurs. These endow the cell with its tissue-specific properties.

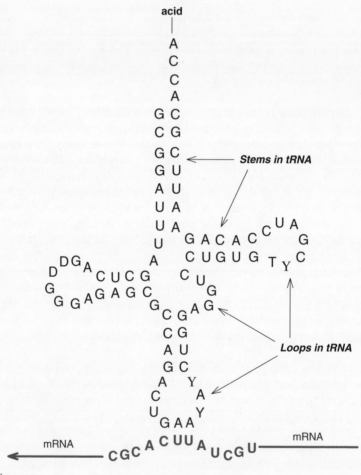

Figure 15.1
The linkage between genotype and phenotype. A codon (←CUU←) in mRNA is "kissed" by an anticodon (→GAA→) in a loop of a tRNA. The particular amino acid attached to that tRNA will be transferred to a growing protein sequence. The two RNA types then dissociate, and another tRNA with a different amino acid attached, seeks out the downstream codon (←GCA←). Note that in RNA the base U replaces the base T (found in the corresponding DNA sequence). Like T, U pairs with A, but also pairs weakly with G.

The processes within a cell which give it the attribute "live" are based on interactions between molecules (proteins, nucleic acids, and other molecular species, both large and small). For example, interactions of specific tRNA molecules with specific proteins, are required to load the tRNAs with appropriate amino acids for protein synthesis. Subsequently, interactions occur between tRNA anticodon loops and specific codons on mRNAs. These interactions are relatively weak (non-covalent), as opposed to the strong (covalent) interactions between the atoms which join to form molecules.

Yet from consideration of the volume available and the large number and variety of molecules dissolved in the cytosol, it is apparent that the cytosol is a very *crowded* place (Fulton 1982). This has important functional implications, both *quantitative* and *qualitative* (Forsdyke 1995a). One quantitative implication is that proteins close to the limit of their solubility tend to aggregate. This is frequently seen when proteins are artificially expressed in a foreign cell. The proteins aggregate as highly insoluble "inclusion bodies." Similar intracellular bodies are seen in various pathological states (Russell 1890; Ordway et al. 1997). A qualitative implication is "non-specific distraction."

NON-SPECIFIC DISTRACTION

Imagine you are seeking someone in a crowded place. You scan those around you, then push your way through to a new part of the crowd and scan again. From time to time you may note people who look a bit like the person you are seeking. You look again to check, and then continue your search. Every person who looks a bit like the person you are seeking is a distractor who marginally slows down your rate of search. In future, to speed up the search process it would be nice if someone would line up every one in the room and eliminate *in advance* the potential distractors. In the absence of this *fine-tuning* of the crowd, your search is less efficient.

Although, electron-micrographs show that the crowded cytosol is structured, in practice the molecules are relatively disordered and seek order (the linking of one molecule with another) through a search which requires stereospecific (lock and key-type) recognition. The price of attempting this in a crowded environment is a large number of interactions, of varying degrees of specificity, with many other molecules. Each of these interactions is non-productive in terms of the proper functioning of the cell, and might cause a slowing of the overall search process and a decrease in the efficiency of cell function.

Having arranged the structure of molecule A such that it can recognize the structure of molecule B, the "hand" of evolution might have achieved a significant advance. However, one would expect that concomitant with and following this, there would be an on-going process of counter-adaptation of other molecules of the cytosol. If a molecule C, which had no functional interest in A, happened to have acquired reactivity with A (by virtue of the modification of A promoting its reactivity with B), then a counter-mutation in C which would not interfere with its normal functioning, but would decrease its specificity for A, would be marginally advantageous. It might take many generations for many such marginally advantageous counter-mutations to recombine together to create a genome conferring a significant increase in the efficiency of cell function. We can refer to this as cytosolic fine-tuning.

By the same line of reasoning, it can be seen that a deleterious mutation in B, decreasing its attraction to A, might at the same time confer on it various degrees of attractiveness for molecules D, E and F. This might have no functional advantage, and might even impair, the functioning of D, E and F.

On the other hand, there is a circumstance where incidental reaction with D, E and F might be turned to the cell's advantage. In chapter 18 it is shown how cells would tend to maintain the concentrations of proteins near to the limits of their solubilities. In chapter 19 it is shown how protein aggregation might, rightly or wrongly, mark a protein as "not-self," thus favouring its presentation at the cell surface, where it would trigger cytotoxic T lymphocytes to destroy the cell. If the protein were a virus protein, this should be to the advantage of the organism. T lymphocytes might also destroy cells transformed to cancer cells because of a mutation in a gene (say B) controlling cell proliferation (e.g., because B no longer interacts with A, cell proliferation might be promoted). The mutation might decrease the solubility of protein B, thus marking it as potentially "non-self" (like a virus protein). If reactions of mutated protein B with unmutated "self" proteins D, E and F, also decreased *their* solubility, then the not-self signal, and hence probability of destruction by cytotoxic T cells, would be amplified (Forsdyke 1999d; 2001h). This would be somewhat analogous to amplification of the not-self signal of an extrinsic prion protein by interaction with the corresponding resident ("self") protein (see chapter 14).

POLYMORPHISM

If a virus protein B aggregated and recruited resident self proteins D, E and F, thus allowing amplification of the not-self signal given by the infected cell to cytotoxic T cells, this would be to the advantage of the host and to the disadvantage of the virus. The T cells would destroy the cell and the viruses it contained. It would be to the advantage of the host to maintain the concentrations of D, E and F near to their limits of solubility. It would be to the advantage of the virus to accept mutations in B which decrease its reactivity with D, E and F or, in some way, to decrease their concentrations. This lesson which it had "learned" on one host (that it is disadvantageous to interact with D, E and F), would be to its advantage on the next host. Since viruses replicate more rapidly and produce more progeny than their host, they can more readily adapt in this way than can the host species. However, the phenomenon of polymorphism, a *discontinuity* among otherwise identical host proteins, can make it difficult for a members of a virus species to exploit what they have "learned" in particular members of the host species.

A host species is a target for a virus species to the extent that members of the host species are uniform. The virus species co-evolves with its host species. By the processes considered in Part 2, all life forms on earth appear, not as a continuum, but as the discontinuities we call species. Within a species there is continuity. It is this continuity (uniformity) which confers vulnerability to pathogens with an ability to adapt more rapidly. Species are like non-moving targets – easy to hit. To the extent that a host species can incorporate discontinuities, without impairing its species integrity, then it has the potential to counter the high adaptability of pathogens. A virus will not be able to exploit what it has "learned" in one host species member, when attacking another host species member.

A polymorphic protein is one which tends to differ among individuals of a species, while retaining its essential functions. Amino acids required for the latter would tend to be conserved both within species and between species. Essential functions might involve amino acids located at the catalytic centre, in the case of an enzyme, and at the protein surface in the case of proteins which are required to interact with other molecules. Amino acids responsible for polymorphism, by definition, do not tend to be conserved, and are often located at the protein surface. Thus, a protein A_1 might be able to mutate to forms $A_2, A_3, A_4, ...$ A_N. In a diploid species, the alleles in one individual might be $A_1 A_3$. In another individual the alleles might be $A_1 A_4$. In another individual the alleles might be $A_2 A_4$. There is a *within-species molecular discontinuity*. Not surprisingly, the most polymorphic loci in vertebrates are the major histocompatibility (MHC) loci, which are involved in immune defences and encode cell surface proteins.

It is not difficult to envision a scenario by which MHC polymorphism arose (Forsdyke 1991). When viruses leave their host cell they often encapsulate themselves in host membrane components. To the extent that these components are uniform from host to host, there is no problem for the virus. However, if there is MHC polymorphism, the membrane from a former host (say $A_1 A_3$) may differ from the membrane of the next host (say $A_1 A_4$). Although A_1 is identical, A_3 is foreign to the new host, and will be the target for immune attack.

Similar considerations can be applied in the case of intracellular host proteins. As indicated above, viral protein B might recruit into aggregation complexes the host proteins D, E and F, in a particular host, thus facilitating host defences. The virus would be unlikely to "learn" from this when attacking its next host if D, E and F were polymorphic to the extent that the versions in the next host, by virtue of their molecular properties, were less likely to interact with B.

That such polymorphism might be to an extent sufficient to confer an adaptive advantage on the host is suggested by the remarkable observation that cancer antigens recognized by cytotoxic T cells are often neither cancer-specific, nor cancer type-specific. Rather, they are specific to the *individual* with cancer. Furthermore, the mutated oncogene protein which started the cancer (say B) is not necessarily the antigenic protein (Srivastava et al. 1998). Thus, in one individual with a cancer the antigens are D, E and F. In another individual with the cancer, the antigens are G, H and I. In another individual with the cancer the antigens are J, K and L (Forsdyke 1999d, 2001h).

SUMMARY

Secondary adaptations, proceeding at a much slower pace than primary evolutionary adaptations, can be viewed as on-going *fine-tuning* allowing gene-products to coexist in a common, crowded cytosol, without interacting. Polymorphisms of gene products are within-species discontinuities compatible with species integrity, which convert species members into "moving targets" thus countering the ability of pathogens to pre-adapt.

16 Fine-Tuning of RNAs

"Chromosomes ... carry out a variety of discrete functions. Each of these functions places informational constraints upon our chromosomes. In a sense, these constraints represent distinct and sometimes overlapping languages. Deciphering some of these languages, and others that we do not have the imagination to envision, will require increasingly powerful computational and mathematical tools."

Hood et al. (1995)

To the several forms of DNA information discussed so far (chapter 14), can be added forms which arose from selective pressures to avoid interactions between RNAs. That the cytosol is not effectively structured to prevent unwanted chance interactions between molecules is evident from the ease with which it is possible to interfere experimentally with cellular processes by injecting or expressing molecules in cells. An "antisense" RNA introduced into a cell can interact with the corresponding complementary "sense" RNA and block its function (Izant and Weintraub 1984). Thus mRNAs are available for RNA–RNA interactions, with the potential to form double-stranded RNA (Melton 1985). If an RNA molecule "wants" to move from one part of a cell to another it would have to cope with a multiplicity of weak distracting interactions with other RNAs. It would be nice if the cytosol were *fine-tuned* in advance to create a "highway" such that RNAs which wanted to go one way could "drive" on one side of the road, and RNAs which wanted to go the other could "drive" on the other.

In addition to the methods discussed in the previous chapter, one way protein molecules might evade distracting interactions is by possession of a similar charge. Molecules with the same charge (positive or negative) would tend to repel each other. If a majority of molecules were similarly charged, there might be a selection pressure favouring retention of this charge and a decrease in the number of oppositely charged molecules. Indeed, most cytosolic proteins are negatively charged, and it is found that the evolutionary rate of change in charge is much less than predicted from the observed rate of accepted amino acid mutations (McConkey

1982). This implies a selection pressure to conserve the negative charge. That a similar principle might also apply to RNA molecules is the subject of this chapter. Although there is no specific "highway," RNA molecules appear to avoid unnecessary interactions by "choosing" between two alternative – purines or pyrimidines.

PURINE-LOADING OF RNA

One of the most important cellular processes is protein synthesis. For this, each tRNA molecule has to discriminate among more than twenty codon sequences in mRNA. In simple terms, assuming random interactions, there is less than a 1/20th chance of encountering the correct anticodon at the first interaction. Yet the search process is incredibly efficient and protein synthesis occurs very rapidly. To achieve this, the physico-chemical state of the cytosol (salt concentration, degree of acidity, concentration of macromolecules) must have been fine-tuned to allow the appropriate search interactions to proceed at great speed. The cytosol is "crowded," a condition known to speed up search interactions leading to hybridization between complementary nucleic acids (e.g., sense – antisense interactions). Furthermore, through evolutionary fine-tuning, molecules capable as acting as distractors in this process should either have been eliminated, or have had their concentrations considerably decreased.

The critical search event between complementary RNA molecules is likely to resemble the reversible loop – loop "kissing" interactions of Tomizawa (1984) which were described in chapter 9. These are between relatively short RNA segments at the tips of stem-loop structures. There are many such short segments, not only at the tips of stem-loops in tRNAs, but also at the tips of stem-loops in other single-stranded RNA species. Irrespective of the details of the mechanism by which a tRNA interacts with its corresponding anticodon on mRNA, it follows from the fact of the existence of a large and varied mRNA population within the cytosol, that the availability of mRNAs for interaction with tRNAs might be severely compromised by chance interactions between mRNA molecules *themselves*. One way mRNAs appear to avoid this distraction is by "purine-loading" their loops.

The story of purine-loading dates back to observations made on various micro-organisms in the laboratories of Szybalski and Chargaff in the 1960s. These organisms have genes compactly arranged in linear order, some being transcribed one way (say to the left), and some being transcribed the other way (to the right). The two complementary strands of transcribed duplex DNA can be considered as consisting of a strand with the same sequence as the mRNA (mRNA synonymous strand), and a strand which acts as a template for the enzyme (RNA polymerase) which copies DNA information into RNA information (template strand). If a gene is transcribed to the left, then the "top" strand is the template strand and the bottom strand is the mRNA synonymous strand. If a gene is transcribed to the right, then the "top" strand is the mRNA synonymous strand, and the bottom strand is the template strand.

Among Chargaff's "rules" on the base composition of DNA (Forsdyke and Mortimer 2000) is his "cluster rule" (Chargaff 1963): "Another consequence of our studies on deoxyribonucleic acids of animal and plant origin is the conclusion that at least 60% of the pyrimidines occur as oligonucleotide tracts containing three or more pyrimidines in a row; and a corresponding statement must, owing to the equality relationship [between the two complementary strands of the DNA duplex], apply also to the purines."

Extending this, Szybalski and his colleagues (1966) found that the mRNA template strand of DNA is characterized by clusters of pyrimidine bases (C or T), and the corresponding mRNA synonymous strand of DNA is characterized by clusters of purine bases (A or G). Meanwhile, Chargaff and his colleagues (e.g., Karkas et al. 1968) had found that Chargaff's first parity rule (A%=T%, C%=G%), which applies to duplex DNA, also applies, *to a close approximation*, to single strands of DNA (Chargaff's second parity rule; see chapter 9).

Combining Szybalski's observation with Chargaff's second parity rule, it follows that the bases in the base clusters (say A) may be balanced by an equal number of dispersed complementary bases (Ts to match As in the same strand). Alternatively, "*to a close approximation*" may mean that there are slight deviations from Chargaff's second parity rule in favour of the clustered bases. Observations from various laboratories show the latter alternative to apply to a wide range of organisms (Smithies et al. 1981; Dang et al. 1998; Bell et al. 1998; Bell and Forsdyke 1999a,b). Indeed, deviations from Chargaff's second parity rule in a strand of DNA are usually predictable if one knows the direction of transcription. Conversely, the deviations can be used to predict transcription direction in an unknown gene ("Szybalski's transcription direction rule"; Figure 16.1).

How did these small deviations from Chargaff's second parity rule arise? If mRNA synonymous DNA strands have purine clusters, it follows that the corresponding mRNAs have purine clusters. Within the cytosol mRNAs are likely to adopt the energetically most-favourable secondary (stem-loop) structures. In the stems, by definition, Chargaff's second parity rule is closely followed (since the stems are formed by complementary base pairs). If a purine cluster were in a stem, there would have to be a corresponding pyrimidine cluster close by in the same strand. Since this does not occur, it is most likely that purine clusters are in the

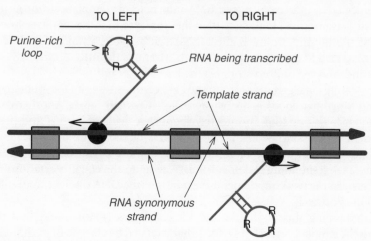

Figure 16.1
Purine-loading. Duplex DNA (long parallel left- and right-pointing thick arrows) is transcribed into RNA by RNA polymerases (black balls) which use either upper or lower strands (small horizontal arrows indicating direction of transcription). Boxes represent non-transcribed DNA between genes (intergenic). The purine-loading of loops in RNAs (R is a generic purine) is reflected in deviations from Chargaff's second parity rule, so that one can tell from the local base composition of one strand in which direction transcription is likely to occur (Szybalsky's transcription direction rule).

loops, as is observed (Bell and Forsdyke 1999b). Thus, the majority of cytosolic mRNAs have purine-rich loops. Since purines do not base-pair strongly with purines, this militates against distracting loop–loop interactions between mRNAs. It is as if RNAs have elected to drive only on the "purine" side of the road. The mRNA synonymous strands of DNA have been under pressure to accept mutations to a purine (A or G), rather than to a pyrimidine (C or T), in regions likely to become loops in the mRNA transcribed from DNA.

MULTIPLE EVOLUTIONARY PRESSURES ON DNA

We have now identified *another* evolutionary pressure on the sequence of protein-encoding genes. Already mentioned pressures are stem-loop pressure (chapter 9), (C+G)% pressure (chapter 10), protein pressure (the pressure to encode a particular amino acid sequence), and the pressure to fine-tune proteins (to avoid non-productive interactions and, through polymorphism, engender productive interactions with pathogen proteins; chapter 15). To these we add purine pressure (the pressure for loops to be enriched in purines). We have also identified *adaptive* forces which are likely to have generated the pressures. Some of these as manifest at the levels of mRNA and the genome are summarized in Figure 16.2 and Table 16.1.

It should be noted that distraction requires only loop–loop interactions involving a few contiguous bases per interaction. If such interactions should progress to the formation of longer segments of double stranded RNA ("consummation"; Figure 9.3), then the distraction would be more profound. It may be that, because of other constraints, it was not possible in the course of evolution to purine-load the loop regions of a particular mRNA to an extent sufficient to prevent "kissing" interactions with similar, but not identical, self mRNAs in the cytosol. The possibility of progression to form longer segments would then arise. This might adversely affect the cell (see chapter 17), so the corresponding mRNAs would have been under pressure to accept further mutations to prevent the consummation of the initial loop–loop kissing interactions.

The various pressures on nucleic acids are not independent. In AT-rich (low percentage C+G) organisms, mRNAs tend to have A-rich clusters. In

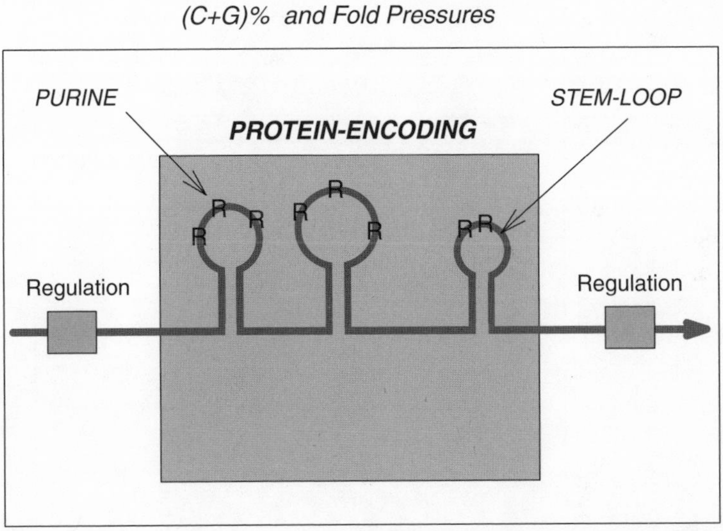

Figure 16.2
Summary of different types of information in mRNA. Messenger RNA (horizontal arrow with stem-loops) must accommodate to (i) genome (C+G)% pressure, (ii) genome fold (stem-loop) pressure (iii) protein-encoding pressure, (iv) pressure to load purines (R) in loops, and (v) pressures for regulation.

Table 16.1
Postulated evolutionary processes leading to multiple levels of information in genomes

Environmental selected factors	Selection for mutations which:	Primary effect on DNA function	Biological result	Observed features of modern DNA
Classical phenotypic selective factors	Change encoded proteins	None	Change in classical phenotype	Usually change in a base pair in accordance with Chargaff's first parity rule
Competitors with more efficient translation, and intracellular pathogens	Purine-load RNAs	None	Efficient translation with no "self" double strand RNA	Chargaff's cluster rule and Szybalski's transcription direction rule
Mutagens	Promote DNA stem-loop potential	Within-species meiotic recombination promoted	Change in genome phenotype for DNA repair	Chargaff's second parity rule
Recombinationally "not-self" sexual partners	Impair homology search between DNAs of species members whose sequences are diverging	Meiotic recombination is impaired	Change in genome phenotype for speciation	Chargaff's (C + G)% rule

CG-rich (high percentage C+G) organisms mRNA tend to have G-rich clusters. However, some organisms have both A-rich and G-rich clusters (Lao and Forsdyke 2000a). With so much information to be carried it is tempting to discount the postulate that many mutations are non-adaptive or "neutral" (see chapters 11 and 13). A single base change (mutation) might affect a protein sequence, or the regulation of gene expression, or local stem-loop potential, or molecular interactions. Many mutations would have to accumulate to affect generally the dispersed, genome-wide, parameter (C+G)%. A gain of a C or G here, might be cancelled by the loss of a C or G there, so that the total (C+G)% in a DNA segment would not easily change. However, an extremely small change in (C+G)% should suffice to affect stem-loop potential, and hence the "reprotype" (see chapter 11 and Bull et al. 1998).

Mutations primarily affecting the phenotype, are referred to here as "phenotypic," to distinguish them from mutations which affect the "genome phenotype" (chapters 10 to 12). Some mutations of the genome phenotype are referred to as "reprotypic," to denote a primary impact on stem-loop potential (and hence on meiotic pairing and reproduction). Such reprotypic mutations can sometimes dominate the protein-encoding function so that less than optimal proteins are produced (see chapter 11). Mutations which enhance purine-loading are primarily phenotypic affecting translation efficiency and, as we shall see in the next chapter, intracellular self/not-self discrimination (Table 16.1); yet even these can sometimes affect the amino acid composition of proteins (and hence, potentially, can affect protein function; Rocha et al. 1999; Lao and Forsdyke 2000a). If protein function is affected then the phenotype may be affected. Thus, any phenotypic character which is not under strict selection has the potential to act as an indicator of an organism's inner constitution, its purine-loading proclivity, or its reprotype. If the goal is reproductive success then there may be something in the adage that one's face is one's fortune.

SUMMARY

Selection pressures for negative charge tend to prevent unnecessary protein – protein interactions. Similarly, the purine-loading of loops in stem-loop structures militates against RNA–RNA (loop–loop) "kissing" interactions which precede the formation of double-stranded RNA. Purine-loading biases the base composition of the mRNA-synonymous strands of DNA, thus allowing prediction of transcription direction in uncharted DNA. Multiple evolutionary pressures on DNA make the neutrality of mutations unlikely.

17 RNAs Driving on the Wrong Side

"If, on Parnassus' top you sit,
You rarely bite, are always bit:
Each poet of inferior size
On you shall rail and criticize;
And strive to tear you limb from limb,
While others do as much for him.
The vermin only tease and pinch
Their foes superior by an inch.
So, naturalist observe, a flea
Hath smaller fleas that on him prey;
And these have smaller yet to bite 'em.
And so proceed *ad infinitum.*
Thus every poet of his kind,
Is bit by him that comes behind."

Jonathan Swift ("On Poetry," 1733)

Conventional natural selection promotes within-species adaptation. The "physiological selection" of Romanes, by modifying DNA to generate exclusivity of recombination, allows divergence (Parts 1 and 2). After divergence, each of the resulting species now becomes part of the environment of the other. Further divergence creates more species and a more elaborate environment. Eventually species appear whose major evolutionary strategy relates to other species. Furthermore, species both diverge and converge. The convergence may be either mutually advantageous (symbiotic), or predatory. The interaction may be at the level of individual whole organisms (lion eats horse), or the interaction may involve a member of one species entering the body of another. Once within the body, the interaction may be primarily extracellular (e.g., most bacteria), or primarily intracellular (e.g., viruses).

When pre-zygotic isolating factors emerged to replace post-zygotic factors, the $(C+G)\%$ which initially made a genome anti-recombinogenic (post-zygotic reproductive isolation) could then have begun to deviate randomly or in response to other pressures. In complex multicellular organisms we find genomes partitioned into sectors which may diverge in $(C+G)\%$ (Bernardi 1993), and this allows the setting up of *within-genome* recombination barriers (see chapter 11). However, such (C+G)% diversity provides more opportunities for recombination with a range of viruses each with a characteristic (C+G) percentage. One outcome is that the ecological niches for a virus within the intracellular environment include the genome of the host cell. Certain viruses have adopted latency strategies, appearing to recombine selectively with sectors of the genome which have a similar (C+G)% as themselves (Rynditch et al. 1998).

In the latent state viruses are quiescent. When it is propitious to do so, viruses can "awaken," and new virus particles can be assembled. In the latter case, viral genomic information is transcribed to generate viral mRNAs which use the same translation system as host mRNAs. Thus, we can ask, should a virus be "polite" and purine-load its RNAs to match host RNAs? In this case both viral and host RNAs could "drive on the same side of the road" and avoid "kissing" interactions. Or, should a virus exploit the fact that host RNAs are purine-loaded, as part of its strategy to adapt and perhaps overwhelm the host cell for its own purposes?

HUMAN T-CELL LEUKAEMIA VIRUS IS IMPOLITE

It was noted (chapter 10) that (C+G)% pressure was likely to have driven apart the (C+G) percentages of various viruses which have the potential to occupy the same host cell (incomplete "pre-zygotic" isolation). Thus the AIDS virus (HIV-1) is AT-rich, whereas human T-cell leukaemia virus (HTLV-1) is CG-rich. These genomes are both transcribed one way (by convention to the "right"). Consistent with the purine-loading rule, for HIV-1 the deviation from Chargaff's second parity rule favours A (e.g., A>T). The expectation would be that for HTLV-1 G>C. However, the opposite holds, namely C>G. Thus, whereas the mRNAs of HIV-1 can be construed as politely "driving" on the purine side of the road like the majority of the mRNAs of their host, the mRNAs of HTLV-1 invite "collisions" since the C-rich loops of viral mRNAs should complement G-rich host mRNAs.

It has been suggested that this difference between HIV-1 and HTLV-1 relates to a difference in their respective strategies or life-styles (Cristillo et al. 1998, 2001). Both viruses have exploited the option of submerging (covalently integrating) within the host genome and lying dormant for a period. When host DNA replicates, virus DNA is also automatically replicated. When in the latent state, transcription from the viral genome is absent, so that viral mRNAs would not be expressed in the cytosol, and so would not interact with ("kiss") host mRNAs. However there is a much *greater* commitment of HTLV-1 to the latency strategy. In contrast to individuals infected with HIV-1 where the disease usually progresses to a fatal outcome in less than a decade, most individuals infected with HTLV-1 remain asymptomatic and live normal lives. Furthermore, HTLV-1 seems able to transfer between individuals when integrated in host DNA *within intact cells*. A person usually acquires HTLV-1 because *cells* containing latent virus are transferred to that person from another individual, whereas a person acquires HIV-1 because virus particles themselves are transferred. Experimentally, naked HTLV-1 virus particles show low infectivity, perhaps because when first in the cytosol viral RNA has to run the gauntlet of interactions with an excess of host RNAs "driving on the normal (purine) side of the road." Why then is HTLV-1 not more "polite"?

When HTLV-1 "decides" to abandon the latent state it transcribes RNAs and "shows its colours." Its aim is to produce many copies of itself and transfer these to other cells and eventually to other hosts. Perhaps when triggered to move from the latent state to one of rapid productive cytolysis (the shedding of viruses accompanied by death of the host cell), the viruses transcribe RNAs which, when released from the nucleus, suddenly flood the cytosol with RNAs "driving on the wrong (the pyrimidine) side of the road." The multiplicity of distracting loop–loop interactions (viral pyrimidine-loaded loops interacting with host purine-loaded loops) might "snarl traffic" and impair the defences of the host cell. This could be of adaptive value to the virus. Studies on other viruses with latency life-styles, namely members of the Herpes group, support this idea.

THE "EBNA-1 ONLY PROGRAM" REQUIRES POLITENESS

Other virus pairs with extreme (C+G)% divergences are in the Herpes family (see chapter 10). *Herpes simplex*-related viruses permanently infect many individuals in their host species, who are often asymptomatic. One of the γ-herpesvirus group, Epstein-Barr virus (EBV; (C+G)%=60) infects lymphocytes (B-lymphocytes) and sometimes this results in the disease known as glandular fever (infectious mononucleosis). Most of us are infected with EBV by the time we reach adult life. The virus is also associated with a tumour known as Burkitt's lymphoma. In contrast, another member of the γ-herpesvirus group, *Herpes virus saimiri* (HVS), which infects T-lymphocytes, is AT-rich ((C+G)%=35). The latter is "polite" (A>T), but EBV is impolite (C>G). Once again, this appears to be associated with the general strategy of latency. EBV lies very low. For much of the time EBV genes are turned off, so the fact that its mRNAs would have C-rich loops is irrelevant.

Intriguingly, a treasured "exception" appears to "prove the rule." Unlike HTLV-1 which does not transcribe during latency, EBV appears to be capable of varying degrees of slumber, and some of its genes (needed to regulate the latent state) are transcribed during latency. This is probably because, unlike HTLV-1, EBV does not integrate into the host genome. It exists as an independent DNA unit (episome), which has to regulate its own rate of replication. The key latency-associated transcript encodes a protein know as EBNA-1 (Epstein-Barr nuclear antigen-1). Whereas the neighbouring genes (turned off during latency) follow the impolite C>G pattern when transcription is to the right, and G>C when transcription is to the left, the rightward-transcribing gene encoding EBNA-1 protein politely follows the *usual* rule (G>C; A>T), and very dramatically so (Figure 17.1). The latter gene is the *only* viral gene expressed in one type of EBV latency which has been called the "EBNA-1 only program" (Thorley-Lawson et al. 1996). Its politeness might relate to the fact that the host's cytotoxic T lymphocyte response is not provoked (Levitskaya et al. 1995), as will be discussed next.

Gly-Ala REGION FOR PURINE-LOADING

The rightward-transcribing gene encoding EBNA-1 should have accepted mutations which increase its purine content (so that its mRNA would follow the normal rule). But what if this were not possible without disrupting protein functional domains? The gene might then be obliged to increase its content of purine-rich codons in inter-domain regions, thus expanding such regions. This provides a possible explanation for a curious feature of the EBNA-1 protein.

Like those of other members of the herpesvirus family, and of viruses in general, the EBV genome is very compact with little intergenic DNA; this suggests an evolutionary selection pressure to eliminate non-functional sequences. However, in the *middle* of the EBNA-1 protein there is a long section of *variable* length in which the two amino acids, glycine and alanine, are repeated (Figure 17.2). The sequence reads Gly,Ala,Gly,Ala,Gly,Ala,Gly,Ala, ... The simplicity of the sequence suggests no subtlety of function. The observed length variations do not seem to affect function. Indeed, the entire segment can be removed from the protein experimentally, without affecting known functions (Yates and Camiolo 1988; Summers et al. 1997). Since the compact virus genome has been under selection pressure to rid itself of unnecessary sequence information, the non-functional simple sequence segment should *not* be present. Presumably it exists because it, *or something obligatorily correlated with it*, is of adaptive value to the virus *under some circumstance*.

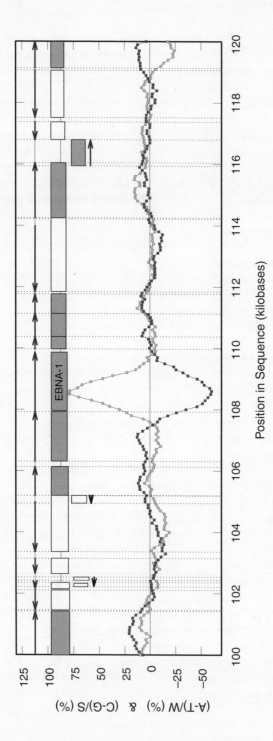

Figure 17.1

Gene expression in Epstein-Barr virus. In contrast to most of its other genes, a gene in Epstein-Barr virus which is expressed during latency (the EBNA-1-encoding gene), has purine-loaded mRNA. Genes are shown as grey or white boxes with arrows indicating direction of transcription. The relative purine-richness for C and G is shown by the squares (e.g., below zero when G is in excess). Similarly, circles are above zero when A is in excess (i.e., A>T).

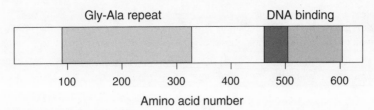

Figure 17.2
Domains of the EBNA-1 protein of Epstein-Barr virus. Removal of the Gly-Ala repeat does not impair known functions. Removal of other domains impairs functions such as DNA replication, protein dimerization, plasmid maintenance, and enhancer activation.

Similar simple sequence segments occur in a variety of other latency-associated genes in members of the herpesvirus family. Explanations are offered in terms of adaptations at the *protein* level (Karlin et al. 1988, 1990; Karlin 1995; Levitskaya et al. 1995). In this case there would have been no obligation to purine-load the mRNA, and pyrimidines might have been well represented. The EBNA-1 simple sequence repeat of glycines and alanines implies that the mRNA encoding the EBNA-1 protein (and the corresponding gene) has a long region containing consecutive codons of glycine (general formula GGN), and alanine (GCN, where N refers to any of the 4 possible bases). In their generic forms, these codons are either purine rich (GGN), or purine neutral (GCN). Table 17.1 shows that choices of third bases (N) in these codons are almost exclusively *purines*. The codon bias in favour of these purines is sufficient to bring the gene into purine excess (G>C), so that the mRNA is polite and follows the normal transcription direction rule.

Although not interpreted as such, the evidence of Klein and coworkers (Levitskaya et al. 1995; Mukherjee et al. 1998) is consistent with the Gly-Ala region being a device for purine-loading a foreign mRNA ("non-self") to make it appear like "self" (host mRNA). If selection had been acting at the protein level to conserve the Gly-Ala region there should not have been such extreme codon bias in favour of purines. In several other members of the herpesvirus family there are similar purine biases in long (>100 amino acids) simple sequence-encoding regions (Cristillo et al. 1998, 2001).

HYPERCHARGE RUNS AS A GENERAL STRATEGY

The use of simple sequence to purine-load mRNAs means that, at the protein level, the simple sequences often contain runs of charged amino acids (e.g., Asp, Glu, Lys) which happen to have purine-rich codons ("hypercharge runs"; Karlin 1995). To compensate for a tendency of its protein

Table 17.1
Codon usage of the glycine-alanine simple sequence repeat region of the EBNA-1 protein of Epstein-Barr virus

Codons	Complete protein*	Less the simple sequence	Simple sequence alone	Human average (%)
Gly GGG	63	11	52	23
GGA	144	43	101	27
GGT	25	24	1	18
GGC	19	19	0	33
Ala GCG	4	3	1	10
GCA	85	2	83	24
GCT	6	6	0	28
GCC	8	8	0	39

*Values are the absolute number of codons in each protein segment

product to provoke autoimmune attack by cytotoxic T cells, a gene might purine-load its mRNAs thus generating long charge-rich alpha-helices which might be irrelevant to the function of the protein itself (Dohlman et al. 1993). In this respect, it is of interest to note the prevalence of charge clusters in proteins implicated in various autoimmune diseases. The clusters do *not* correspond to the antigenic parts of the proteins (Brendel et al. 1991). This suggests that the charge cluster domains may not be the *primary* cause of the diseases, as has been suggested, but may have evolved as a necessary correlate of the purine-loading which occurred in response to the disease-provoking characteristics of *other* domains.

One form of purine-loading involves the repeated use of the codon CAG (which encodes the amino acid glutamine). Unfortunately such repeats predispose single-stranded nucleic acid to adopt stem-loop conformations. DNA at the point of replication appears readily to adopt such conformations with the mechanical consequence that the repeat may rapidly increase in length. The corresponding polyglutamine tract, playing no role in the normal function of the encoded protein (Ordway et al. 1997), is then increased in length to such an extent that the protein becomes insoluble and aggregates (as in the prion diseases mentioned in chapter 14). Neurological diseases associated with this "triplet expansion" include Huntington's disease (Green and Wang 1994).

MESSENGER RNAS AS "ANTIBODIES"

So why is EBNA-1 mRNA so polite? The first line of defence against a foreign *extracellular* agent (bearing antigenic proteins at its surface) are the protein antibodies which are synthesized by B-lymphocytes. These antibodies cover a wide range of specificities so that for any particular antigen there are likely to be at least a few antibody molecules which can bind the antigen with sufficient affinity to mark the extracellular agent as foreign (not-self). An individual mature B-lymphocyte has the potential to make antibody molecules of just one specificity. If the antigen happens to bind to the antibody-like receptor of that particular B-lymphocyte then the latter is likely to proliferate and secrete more antibody molecules of the same specificity. In this way the antigen "selects" B-lymphocytes able to form antibodies which can bind to it.

If the extracellular agent is a virus it must evade these antibody "missiles" if it is to achieve its goal of penetrating an appropriate target cell. Far from stormy extracellular seas it seeks safe haven in tranquil intracellular waters. Here the antibodies cannot penetrate. Even if they could, the virus has left its highly antigenic outer coat behind and initially exists within cells just as a foreign nucleic acid molecule. Eventually it will program the host cell to make more viral coat proteins, which might then be recognized by the cell as foreign *if there were an appropriate protein-orientated intracellular defence system* (see chapter 19). However, an earlier opportunity to alert intracellular defences would arise if it were possible to recognize viral nucleic acid as not-self.

Distracted by the *messenger* role of mRNA molecules, we may fail to note that the diverse spectrum of *intracellular* mRNA species, like the diverse spectrum of *extracellular* antibodies, constitutes a repertoire of specificities with the potential to react with complementary sections of non-self RNA "antigens." If EBNA-1 mRNA ("sense") in latent EBV-infected cells were not "polite" (to avoiding "kissing"), then it is possible that a few host mRNAs would have a sufficient degree of complementarity ("antisense") to progress beyond kissing interactions. Indeed, individuals with mutations in RNAs promoting such consummation could differentially survive since double-stranded RNA molecules alert host defence systems (Fire 1999; Hamilton and Baulcombe 1999).

Current evidence suggests a minimum of two helical turns (at least 22 base-pairs) would be necessary for this (Robertson and Mathews 1996; Tian et al. 2000), and the specificity repertoire of host mRNAs alone would not approach the 4^{22} possible combinations of four bases. However, each virus would itself present a repertoire of 22-base RNA target "antigens." There is also a low level of host read-through transcription of (failure to terminate RNA synthesis from) both genes and repetitive elements in intergenic DNA. So the host RNA "antibody" repertoire could be supplimented with non-translated RNA transcripts from extragenic DNA. This provides a role both for "junk DNA" and for the "selfish" repetitive elements it contains. To this extent we could, after all, be at the "centre" of our own DNA (see chapter 14).

Given that a length of double-stranded RNA sufficient for an alarm signal had been generated, there would then be increased synthesis of proteins which assist defences against viruses (interferons and MHC proteins; Sen and Lengyel 1994). Thus, through its politeness, EBNA-1 mRNA would avoid increasing MHC protein expression in this way, and the MHC-dependent cytotoxic T-cell response would be impeded. Hence maintenance of EBV in the latent state would be uninterrupted (Cristillo et al. 2001). If this were a general trend in latent viruses, then genes expressed as proteins during latency could differ in their codon usage patterns from genes expressed during the productive phase of infection, as is indeed found (Karlin et al. 1990). It is of interest that vaccinia virus, which *productively* infects (i.e., is not latent within) its host cell, has purine-rich loops in its mRNAs, like those of its host, while mRNAs of *Herpes simplex* virus (*latent* in the infected cell), have pyrimidine-rich loops.

These considerations suggest that, not only a foreign protein, but also a foreign nucleic acid, can provide appropriate non-self signalling for alerting the immune system. Although the initial pressure for the evolution of purine-loading may have been simply to decrease distracting RNA-RNA interactions, it can be seen that this made it possible for the evolution of defences based on recognition that unacceptable lengths of double-stranded RNA had been formed. Protein aggregates are a major feature of the prion and triplet aggregation diseases, but evidence that the aggregates per se are responsible for cytotoxic effects is not clear. Both prion and triplet expansion disease mRNAs have repeats capable of forming lengths of double-stranded RNAs sufficient to alert host defences (Lao and Forsdyke 2000a; Forsdyke 2000c). Thus, it is possible that the cytotoxic effects observed in these diseases are triggered by double-strand RNAs and mediated by the immune system.

SUMMARY

Over evolutionary time, viral species diverge from and converge with cell-based biological species. Fine-tuning, as described in the previous chapter, helps a host cell detect and discriminate against a virus. The AIDS virus is "polite" and has tuned its RNA to match those of its host. Viruses with a higher commitment to latency are infrequently expressed in the cytosol, and are not so accommodating. However, latent viruses which synthesize a few latency-associated RNAs may adapt these particular RNAs for politeness. This may explain a mysterious simple-sequence element found in the major latency protein of Epstein-Barr virus. Intracellular RNAs may relate to foreign RNAs in the same way extracellular antibodies relate to foreign proteins (antigens). The resulting double-stranded RNAs would trigger host defences against viruses, just as antigen-antibody complexes can trigger inflammatory responses.

18 Protein Concentration and Genetic Dominance

> "Population geneticists have yet to remove from the theoretical framework many of the basic differences of approach already visible in 1932 in the work of Fisher, Haldane and Wright. For example, each proposed a model for the evolution of dominance and these have yet to be authoritatively reconciled."
>
> W.B. Provine (1971 ch.5, 177)

While not a reconciliation of theories of the evolution of genetic dominance, there now seems to be a stronger case for the "dose-response" or "physiological" theory of one party (Haldane 1930; Muller 1932; Wright 1977 ch.15, 513), than that of the other (Fisher 1931). In 1980 Kacser and Burns presented a detailed enzyme kinetic account of the dose-response theory. We will here consider a simpler version implicating a strange class of proteins, known as "heat-shock" proteins, in the evolution of dominance (Forsdyke 1994a).

DOMINANT AND RECESSIVE GENES

In a diploid organism there are two sets of chromosomes (homologs), one set carrying paternally-derived genes and the other set containing maternally-derived genes. The paternal and maternal copies of a gene are potential alternative versions of that gene ("alleles"). As mentioned in chapter 2, sometimes the alleles are identical and the organism is said to be *genotypically homozygous* for that gene. There may be small differences between the alleles which do not affect phenotype, in which case the organism would be *phenotypically homozygous*. If not qualified, the term homozygous in the genetic literature usually means phenotypically homozygous. Mendel (1865) defined genes as dominant if the expression in heterozygotes matches that in the homozygotes, and recessive if the expression in heterozygotes is considerably less than that in homozygotes. When crossing a homozygous tall pea plant with a homozygous small pea plant, the first generation of plants (heterozygotes) were all tall, so that the gene (allele) for tallness is dominant and the gene (allele) for smallness is recessive (assuming just one critical gene is involved). If there is an allele which predominates in natural environments it may be referred to as the wild-type allele.

Mendel noted (Iltis 1932 ch.14, 197) that "hereditary characters and visible characters must be carefully distinguished from each other." Unaware of Mendel's work, Darwin wrote (1875 ch.14, 60): "We can seldom tell what makes [a character in] one race or species prepotent [dominant] over [the corresponding character in] another; but it sometimes depends on the same character being present and visible in one parent, and latent or potentially present in the other." Since a dominant character is defined by its being "present and visible," it is merely to repeat the definition to say that dominance "depends on the same character being present and visible in one parent." The importance of the remark is that Darwin recognized that a character can exist within the organism in some "latent or potentially present" state. Similarly, de Vries (1889, 33) concerning "the hereditary factors, of which the hereditary characters are the visible signs," noted that "frequently the conditions are so unfavourable for some of them that they cannot manifest themselves at all, and so remain latent." Thus, for a recessive character, although there is absence of evidence for the corresponding hereditary factor in the heterozygote, the factor may still be present. Absence of evidence is not evidence of absence.

Bateson (1909b ch.1, 11) concerning tallness and smallness in peas noted: "There must at some stage in the process of germ [gamete] formation be a separation of the two characters, or rather the *ultimate factors* which cause these characters to be developed in the plants. This phenomenon, the dissociation of characters from each other in the course of the formation of the germs, we speak of as *segregation* [Bateson's italics], and the characters which segregate from each other are described as *allelomorphic* [allelic; Bateson's italics], i.e., alternative to each other in the constitution of the gametes." He emphasized (1909b ch. 15, 288): "The notion that a character once appearing in an individual is in danger of obliteration by the intercrossing of that individual with others lacking the character proves to be unreal; because *in so far as* the character depends on [ultimate] factors which segregate [among gametes], no obliteration takes place. The [ultimate] factors are permanent by virtue of their own properties, and their permanence is not affected by crossing."

Regarding homozygotes, Bateson noted (1914, 281): "Since [ultimate] genetic factors are definite things, either present or absent from any *germ cell* [gamete], the [diploid] individual may either be 'pure bred' [homozygous] for any particular factor, or [for] its absence, if he is constituted by the union of two germ cells both possessing or both destitute of that [ultimate] factor." Regarding the dominance or recessiveness of heterozy-gotes carrying hereditary diseases, he noted (1909b ch.12, 232):

> If ... a disease descends through the affected persons, as a dominant, we may feel every confidence that the condition is caused by the operation of a factor or element added to the usual ingredients of the body. In such cases there is something *present*, probably a definite chemical substance, which has the power of producing the affection. ... On the contrary, when the disease is recessive we recognize that its appearance is due to the *absence* of some ingredient which is present in the normal body. So, for example, albinism is almost certainly due to the absence of at least one of the factors, probably a ferment [see chapter 6], which is needed to cause the excretion of the pigment" [all italics are Bateson's].

We find here a distinction between "ultimate factors," which we now know as genes, active "factors" or "ingredients," which we now know best as enzymes ("ferments") or their products, and the observed "characters" of an organism, which we now call its phenotype.

In his experimental work Bateson (1909b ch.4, 76) found it "evidently simpler" to blur some of these distinctions and "imagine that the dom-inant character is due to the *presence* of something which in the case of the recessive is *absent*" [Bateson's italics]. It is evident that he was refer-ring to "active factors," not "ultimate factors" when writing (1909b ch.2, 53-4): "A dominant character is the condition due to the *presence* of a def-inite factor, while the corresponding recessive owes its condition to the *absence* of the same factor. ... The green pea, for instance, owes its reces-sive greenness to the *absence* of the [active] factor which, if present, would turn the colouring matter yellow" [Bateson's italics]. He further noted (1909b ch.4, 76): "In cases where the pure [homozygous] domi-nants are recognizably distinct from the heterozygous dominants [i.e., dominance is incomplete], it must naturally be supposed that two 'doses' of the active factor are required, one from the paternal, and another from the maternal side, in order to produce the full effect."

THE DOSE-RESPONSE THEORY

The dose-response theory states that the quantity of the product of a dominant wild-type allele in a homozygote is so much in excess of the needs of the organism that halving this quantity, as might occur in a het-erozygote containing a wild-type allele and a mutant allele (encoding a non-functional product), will not change the phenotype. Figure 18.1 shows a plot of some quantifiable phenotypic feature against the dose of the product of a gene concerned with that phenotype. At low doses of

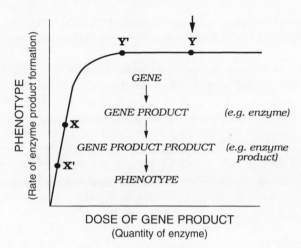

DOSE OF GENE PRODUCT
(Quantity of enzyme)

Figure 18.1
Dose-response curves showing a measure of phenotype plotted against the dose of a
gene product which contributes to that phenotype. The thick vertical arrow indicates
the normal concentration (Y) of a non-rate-limiting gene product in a diploid
homozygous cell. Halving this dose (Y → Y') has no effect on phenotype (i.e., the
concentration is still not rate-limiting). X and X' indicate corresponding points for a
rate-limiting gene product. (Also shown is the hierarchical flow of information from
gene to phenotype, in a typical case.)

gene product (indicated by point X on the curve) the phenotype depends
directly on gene-product quantity. Halving the dose (point X') halves the
phenotypic parameter. At high doses of gene product a plateau is
approached when some other factor becomes limiting (e.g., the availability of substrate A; see later). Under these conditions increasing the quantity of a gene product has no effect on the phenotypic parameter. In the
case of the wild-type homozygote it is postulated that the normal
amount of gene product corresponds to a point well along the plateau of
the curve (indicated by point Y). This creates a "factor of safety"
(Haldane 1930), or "margin of stability and security" (Muller 1932), so
that halving the amount of the product (to point Y'), as would occur in
a heterozygote, has no effect on the phenotypic parameter.

RATE-LIMITING AND NON-RATE-LIMITING
STEPS

In many cases the phenotype will be the result of a series of enzyme catalyzed reactions such as shown in Figure 18.2. Substrate A is converted to
end-product D through a series of intermediates (B, C), catalyzed by

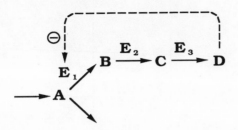

Figure 18.2
A metabolic pathway with feed-back inhibition. A is the substrate of enzyme E_1, which catalyzes the rate-limiting step in the pathway. A is also metabolized by another pathway. B and C are substrates of enzymes E_2 and E_3, which are not rate-limiting. D gives the phenotype and feed-back negative regulation at E_1.

enzymes E_1, E_2, and E_3. The first step of the pathway, the rate-limiting step, is subject to feedback inhibition by D.

Figure 18.3 shows hypothetical *in vivo* substrate dose-response curves for the three steps in the pathway. The vertical arrows indicate the normal substrate concentration existing within cells. In the case of the *rate-limiting* enzyme E_1 the cellular concentration of A must, by definition, correspond with the plateau of the dose-response curve, so that enzyme concentration, and not substrate concentration, is rate-limiting. This would correspond to point X on the plot of reaction rate versus enzyme concentration (Figure 18.1). In the case of the *non-rate-limiting* enzymes E_2 and E_3, the normal substrate concentration corresponds to the ascending limbs of the corresponding substrate dose-response curves (Figure 18.3). There is ample enzyme to accommodate fluctuations in availability of the substrates B and C (the products of E_1 and E_2, respectively). This quantity of enzyme would correspond to point Y on a plot of reaction rate versus enzyme concentration (Figure 18.1). Thus, after a molecule of A has squeezed through the "bottleneck" E_1 to become B, subsequent chemical modifications by E_2 and E_3 do not influence the rate of accumulation of the end product D.

It can be seen that the situation with the non-rate-limiting enzymes (E_2, E_3) corresponds to the "margin-of-safety" scenario. In the case of the rate-limiting enzyme E_1, halving of gene-product concentration (X –> X' in Figure 18.1) would have consequences for the phenotype. However, in general, rate-limiting enzymes are subject to complex controls by products of intermediary metabolism. The most common of these is end-product inhibition (Figure 18.2). A decrease in D in the heterozygote would tend to decrease the inhibition. This would increase the activity

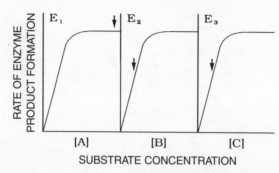

Figure 18.3

Substrate dose-response curves for the rate-limiting enzyme E_1, and the non-rate-limiting enzymes E_2 and E_3, within cells. Rates of formation of products (B, C, and D) are expressed as functions of the concentrations of substrates A, B, and C, respectively. The vertical arrows refer to the normal concentrations of these substrates interacting with the enzymes within living cells.

per molecule of E_1, so that the reaction rate would become the same as in the homozygote. Thus, in this case, feedback inhibition provides another "margin of safety" allowing a heterozygote to maintain the wild-type phenotype.

Some of the factors contributing to the "decision" on the optimum concentration of a given protein within intact cells are summarized in Figure 18.4. There must be a sufficient concentration for function (i) in the homozygote, (ii) in the heterozygote, and (iii) with the collective of other cytosolic proteins to exert an aggregation pressure favouring intermolecular interactions (see chapter 19). At the same time the concentration must not exceed a threshold above which the protein will self-aggregate. Extreme manifestations of this are insoluble "inclusion bodies" within cells, and fibrillar precipitates (e.g., amyloid) outside of cells. The perils of insolubility being so great, it is not surprising that we find certain proteins within cells which function to assist other proteins to maintain conformation. One class of these "molecular chaperones" is that of the heat-shock proteins.

"EXTREME ENVIRONMENTAL DISTURBANCES"

A major problem with what we might now call the margin-of-safety theory, is in determining what selective forces would have created and sustained the margin of safety. Some have argued that this can be explained entirely in metabolic terms (Kacser and Burns 1980). In the case of a rate-limiting enzyme (E_1) the margin could indeed be a simple consequence of

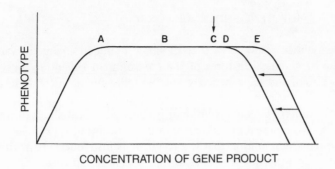

CONCENTRATION OF GENE PRODUCT

Figure 18.4
Factors affecting protein concentration within living cells. The point (C) on the plot of a variable phenotypic character versus gene product (protein) concentration, corresponds to the normal concentration of the protein responsible for that character within a living cell. The phenotypic parameter increases with gene dosage until point A when some other factor (e.g., substrate availability) becomes rate-limiting. The curve then plateaus. B corresponds to the minimum concentration of gene product needed for a "margin of safety," so that heterozygote function is not impaired. E corresponds to the concentration at which the protein would still be soluble if no other proteins were present. Above this concentration the protein would self-aggregate. D corresponds to the concentration at which aggregation would occur in the presence of cytosolic proteins. Horizontal arrows symbolize the aggregation pressure exerted collectively by cytosolic proteins, so moving the descending limb of the dose-response curve to the left.

the evolution of metabolic controls. It is in the case of non-rate-limiting enzymes (most enzymes) that a further explanation must be sought. What sustains the enzyme activity in a wild-type homozygote at point Y rather than at point Y' (Figure 18.1)?

In a population there are both homozygotes and heterozygotes. Muller (1932) thought that evolutionary selection would primarily work through homozygotes. He postulated that "mutations favouring dominance ... have been selected and are maintained, not so much for their specific protection against heterozygosis at the locus in question, but as to provide a margin of stability and security, to insure the organism against weakening or excessive variability of the character by *other and more common influences*, environic [environmental] and probably also genetic." Along similar lines Wright (1977 ch.15, 524) stated: "Because of *occasional extreme environmental disturbances*, a considerable excess [of gene product] is advantageous. This is likely to be so great that the correlated response of the rare heterozygote is also brought fairly close to the asymptote, thus giving a high degree of dominance." The nature of the "occasional extreme environmental disturbances" was not specified.

THE HEAT-SHOCK RESPONSE

The heat-shock response of a cell follows a sudden change in various physical or chemical features of the environment (including infection with a virus), and is particularly notable following an increase in temperature. The response is detected as a rapid *increase* in the intracellular concentrations of a set of evolutionarily conserved "heat-shock proteins," which is accompanied by a *decrease* in the concentrations of many normal proteins. The notion that the response is just concerned with protection against thermal stress has lost ground in recent years.

It has been proposed that what we call the heat-shock response has actually evolved as part of a mechanism for distinguishing the proteins of intracellular pathogens (not-self) from normal intracellular proteins (see chapter 19). The proteins of the crowded cytosol are held to exert a collective pressure *on each other* tending to make individual protein species aggregate when their concentrations exceed their individual solubility limits (Forsdyke 1995a). These concentrations have been fine-tuned over evolutionary time so as not to exceed these limits. Not-self proteins may more readily "trip" an intracellular surveillance system because their concentrations have not been so fine-tuned.

The aggregations, being of a type described as "entropy-driven," are strongly increased by an increase in temperature (Lauffer 1975). The organism exploits this (e.g., fever) to promote the aggregation of the proteins of a foreign pathogen. In this process self proteins might also be aggregated. To decrease this, the concentrations of normal proteins decrease. However, in turn, this decreases the collective pressure exerted by the cytosolic proteins which tends to make the proteins of the pathogen aggregate. To compensate for this, a special set of proteins, the heat-shock proteins, increase in concentration. Some of these have a role in processing fragments of the aggregated proteins, whereas, in the unshocked cell, heat-shock proteins would manifest mainly their chaperone role. Thus, there is a "mode switch" from chaperone-mode to fragment processing-mode (Forsdyke 1999d, 2000c, 2001h).

The main point to be made about the heat-shock response in the present context is that it probably reflects a fundamental process which appeared early in evolution when sets of replicators encased in a membrane (prototypic cells) had to be protected against invasion by foreign replicators (prototypic viruses; Forsdyke 1991). All subsequent evolutionary developments would potentially be influenced by this pre-existing system. The sudden general fall in the concentration of normal self proteins as part of the heat-shock response would severely compromise cell function if there were not a margin of safety regarding function. Thus the heat-shock response would constitute a powerful evolutionary force

Figure 18.5
Herman J. Muller (1890-1967) in 1940.
Photograph by H. Reichenbach.

acting on wild-type homozygotes. This would lead to the general evolution of proteins of a specific activity sufficient to sustain (or facilitate the recovery of) cell function, at a time when protein concentration had fallen. An incidental outcome of this would be that a heterozygote would normally have sufficient gene product so that it would be phenotypically indistinguishable from the wild type.

However, having only one functioning copy of the allele in question, a heterozygote would have sacrificed part of its margin of safety and this might have some impact on viability. Indeed, heterozygotes for lethal mutations sometimes show some reduction in viability compared with the homozygous wild types and may be more temperature sensitive.

DOSAGE COMPENSATION

The hypothesis offers a new way of looking at the problem of X chromosome dosage compensation, which is the subject of the next chapter. This was first described as the process by which the function of the single X chromosome of male fruitflies is made equivalent to the function of both X chromosomes in females (Muller 1948; Figure 18.5). Without dosage compensation, the situation would be formally equivalent to the points Y (in females) and Y' (in males) as shown in Figure 18.1. Muller postulated that these "exceedingly minute" phenotypic differences (differences between the point Y and the point Y' phenotypes) would constitute a sufficient selection pressure for dosage compensation to have evolved. The heat-shock response, in shifting heterozygote (male) gene product concentrations from point Y' to point X (Figure 18.1), would have created a much greater selection force for the evolution of a margin of safety in males. Although this might have been a factor in the evolution of dosage compensation, it is argued in the next chapter that the major factor is probably the need to fine-tune protein *concentrations*, rather than protein functions, to be equal in male and female cells.

SUMMARY

Fisher disagreed with Haldane, Muller, and Wright on genetic dominance. The dispute can be resolved in favour of the dose-response theory of the latter authors. This requires selection for cytosolic protein concentrations such that there is a "factor of safety" against decreases in protein concentration. The "extreme environmental disturbance" invoked by Muller and Wright to explain the evolution of the safety factor, could be the heat-shock response. Collectively, proteins in the crowded cytosol exert an aggregation pressure which tends to limit the solubility of individual proteins. Thus the factor of safety in protein concentration has an upper limit imposed by companion proteins. When this limit is exceeded self-aggregation occurs.

19 Sex Chromosomes

"It is a relief to have some feasible explanation of the various facts, which [explanation] can be given up as soon as any better hypothesis is found."

Charles Darwin 1868 (Darwin and Seward 1903 ch.4, 301)

There are few areas in biology where "the various facts" are more in need of "some feasible explanation" than those of sex-chromosome differentiation and the associated phenomenon of dosage compensation. In many species, males and females are produced in equal quantities. Mendel himself is interpreted as imputing that such equal quantities would be produced if a recessive homozygote were crossed with a heterozygote (Bateson 1909b ch.10, 165). Thus, if red is dominant to white, when a homozygous white (*WW*, producing one type of gamete, *W* and *W*) is crossed with a heterozygous red (*RW*, producing two types of gamete, *R* and *W*), on average equal numbers of red and white progeny should be produced (*RW*, *WW*). If sex were similarly determined, this simple scheme would require that one sex be homozygous and the other sex be heterozygous for alleles of a particular gene.

"REPRODUCTIVE ISOLATION" OF SEX CHROMOSOMES IN ONE SEX

As long as one allele pair were required there would be no reason to regard this process as different from any other genetic process. The chromosome pair containing the gene – let us call them X and Y chromosomes – would be equal in all respects, including size ("homomorphic"). One sex would be the recessive homozygote (the "homogametic" sex) and the other sex would be the heterozygote (the "heterogametic" sex). If there were recombination between the chromosome pair in the heterogametic sex, the gene might switch chromosomes, thus converting an

X chromosome to a Y chromosome and the corresponding Y chromosome to an X chromosome. The *status quo* would be preserved and equal numbers of differentiated gametes would still be produced.

However, if the complexities of sexual differentiation were to require more than one gene – say genes X_1 and X_2 on the X chromosome, and the corresponding allelic genes Y_1 and Y_2 on the Y chromosome – the situation might get more complicated. Recombination might separate the two genes to generate chromosomes (and hence gametes) with genes X_1 and Y_2 together, and genes Y_1 and X_2 together. Sexual differentiation might be impaired. To prevent this happening members of the gene pairs – X_1 and X_2, and Y_1 and Y_2 – would have to be closely linked on the X and Y chromosomes, respectively, and/or the chromosomes would have to develop some local mechanism to prevent recombination (Ohno 1967 ch.2, 16-23). If such anti-recombination activity could not easily be localized, then the activity might spread to involve other genes on the chromosomes; these genes might themselves play no role in sexual differentiation. This seems to be the situation that often prevails. The predominant function, sexual determination, overrules, but does not necessarily eliminate, the functions of other genes on the same chromosome. The chromosomes then are referred to as the sex chromosomes, even though concerned with many functions not related to sex.

But recombination has evolved because it is advantageous, as alluded to earlier (see chapter 8). The benefits of recombination are not to be lightly discarded. We have seen that, for the initiation of speciation, recombination is a hazard; the prevention of recombination between members of different incipient species, or of an incipient species and the parent species, is a fundamental part of the speciation process ("reproductive isolation"; see Part 2). However, sex chromosomes exist *within* a species, and anti-recombination is only beneficial to the extent that it prevents recombination between genes specifically concerned with sexual differentiation. If anti-recombination activity should spread to encompass genes not concerned with sexual differentiation, the latter would lose any benefits recombination might have conferred. To the extent that recombination allows the correction of damaged genes, any non-lethal mutations resulting from such damage would remain uncorrected. There would be no prevention, *within* species, of further changes in sex chromosomes, including additions and deletions of segments ("macromutations") similar to those which occur *between* species as they differentiate (Chandley et al. 1975).

Thus, instead of remaining homomorphic, the sex chromosomes might come to differ in size (heteromorphic). In some species (e.g., certain fish and amphibia) the sex chromosomes remain homomorphic and appear to differ only in the genes affecting sexual differentiation. In other species (e.g., mammals, birds) the sex chromosomes are indeed hetero-morphic. It has been argued that modern heteromorphic sex chromo-somes evolved from homomorphic ancestors (Ohno 1967 ch.2, 16-23).

HALDANE'S RULE

So, in many species the sex of an individual depends on which of two alternative sex chromosomes he/she inherits. The preservation of the anatomical and functional characteristics which differentiate the *sexes* requires "reproductive isolation" (anti-recombination activity) *within* a species, just as the preservation of the anatomical and functional charac-teristics which differentiate allied *species* requires reproductive isolation (anti-recombination activity) *between* species. In the case of sexual dif-ferentiation, the reproductive isolation occurs at the chromosomal level, and is manifest in only *one* of the sexes. Human females (the homoga-metic sex in our species) have two X chromosomes, one inherited from the father and one inherited from the mother. These are essentially iden-tical and behave during meiosis like the non-sex chromosomes (auto-somes).

On the other hand, human males (the heterogametic sex) inherit an X chromosome from their mother and a small Y chromosome from their father. With respect to the regions of these chromosomes containing the genes affecting sexual differentiation there must be the equivalent of the hybrid sterility which affects crosses between allied species (see chapter 4). These regions must not be allowed to recombine (exchange DNA segments) because then the characters conferring sexual identity might lose their association with a specific sex, and sexual differentiation might be lost.

In hybrid sterility, anti-recombination develops between homologous chromosomes. The anti-recombinational effect of $(C+G)\%$ divergence as proposed for incipient speciation (see Part 2), seems applicable, in principle, to the initiation of the anti-recombinational activity required to prevent recombination in the heterogametic sex between genes required for incipient sexual differentiation. A difference is that, in hybrid sterility, the $(C+G)\%$ divergence between previously homologous chromosomes triggers "check-point" controls which impair meiosis and subsequent gametogenesis. Clearly this does not result when the sequences of sex chromosomes diverge *within* a species. Normally game-togenesis is unimpaired in the heterogametic sex.

However, a step towards speciation would have been taken in the het-erogametic sex (e.g., the sex with X and Y chromosomes); regions of the two sex chromosomes would be "reproductively isolated" from each other, allowing preservation of specific types (sexes). The process of incipient speciation requires that (i) sex chromosomes diverge, (ii) autosomes diverge, and (iii) "checkpoints" interfering with meiosis are activated. The homogametic sex has to accomplish all three steps. The heterogametic sex has already accomplished much of the first step. Thus, if divergence of sex chromosomes for sexual differentiation is mechanistically similar to the divergence of autosomes for species differentiation, then sexual differentiation appears as a "way station," *en route* to species differentiation. Indeed, observing that sexual and species differentiations can involve morphological and physiological changes of similar orders, Bateson (1904b) noted:

> Whenever sexual organisms breeding together produce a mixture of forms, there is ... *prima facie* reason to suspect that the mixture is due to differentiation of germs [gametes]. The most familiar case is sex itself. A population consisting of males and females has so many features in common with the differentiating offspring resulting from the segregation of characters among the germ-cells of cross-bred [hybrid] organisms that it is impossible to avoid the suspicion that the two phenomena [sexual differentiation and species differentiation] are similar in causation.

At a time when chromosomal-determination of sex was still poorly understood he further noted (1908, 329-332):

> We may feel fairly sure that the distinction between the sexes depends on the presence in one or other of them of an unpaired factor. ...The results of genetic research are so bewilderingly novel ... . In all the discussions of the stability and fitness of species, who ever contemplated the possibility of a wild species having one of its sexes permanently hybrid [i.e., X and Y chromosomes]? ... Who would have supposed it possible that the pollen cells of a plant could be all of one type, and its egg cells of two types.

This acknowledged studies of his colleague Edith Saunders on certain dioecious plants (plants with two independent sexes) in which the female is the heterogametic sex (i.e., has two types of gamete), and the male is the homogametic sex (i.e., has one type of gamete). By 1922 the chromosomal basis of sexual differentiation was becoming clearer and in his Toronto address to the American Association for the Advancement of Science Bateson commented on the chromosomally-borne "ingredients" responsible for this:

> We have now to admit the further conception that between the male and female sides of the same plant these ingredients may be quite differently apportioned, and that the genetical composition of each may be so distinct

that the systematist might, without extravagance, recognize them as distinct specifically [i.e., mistake the different sexes as distinct species]. If then our plant may give off two distinct forms, why is not that phenomenon a true instance of Darwin's origin of species?

A corollary of this is that incipient speciation, manifest as some *degree* of hybrid *sterility* when "varieties" were crossed, would be predicted to appear at the earliest stage in the heterogametic sex, even in genera with homomorphic sex chromosomes (Figure 19.1; Naveira and Maside, 1998). Indeed, perhaps with guidance from Bateson (see chapter 21), Haldane (1922) noted this is the "rule" for hybrid sterility. Later as differentiation increased in the autosomes, reproductive isolation between the "varieties" (then, by definition, "species") would be complete.

Haldane's "rule" can also be manifest as hybrid *inviability*. A proposed mechanism for the latter (Forsdyke 1995b) predicts interference with the process which is considered later in this chapter – dosage compensation. The mechanism of interference with dosage compensation, which will not be discussed here, would *not* be expected to apply to crosses between "species" (varieties) of genera with homomorphic sex chromosomes, such as mosquitoes of the genus *Aedes*. Indeed, Presgraves and Orr (1998) found for crosses between *Aedes* "species" that Haldane's rule was associated with male hybrid sterility, *not* inviability.

Although the differentiation of sex chromosomes might have initiated by way of differences in $(C+G)\%$ (micromutations), later macromutations appear to have included extensive deletions on one sex chromosome, so that in some species one of the sex chromosomes appears to have degenerated, losing most of its genes (i.e., the organism has become haploid for those genes). One sex, the homogametic sex (female in humans), contains two copies of the normal, undegenerate chromosome (X). The other sex, the heterogametic sex, contains one normal chromosome (X) and one degenerate chromosome (Y). These show sequence similarity only in short ("pseudoautosomal") segments, where recombination can occur. As proposed for speciation (chapter 12), the subsequent macromutations (e.g., deletions in the Y chromosome) might have effectively substituted for the original $(C+G)\%$ divergence in keeping genomic regions concerned with sexual differentiation recombinationally isolated, so that no trace of the original mechanism of divergence might be evident in many modern species.

IMPLICATIONS OF HAPLOIDY IN ONE SEX

Irrespective of the mechanism of its evolution, the frequent haploidy (hemizygosity) of the sex chromosomes in the heterogametic sex is a fact,

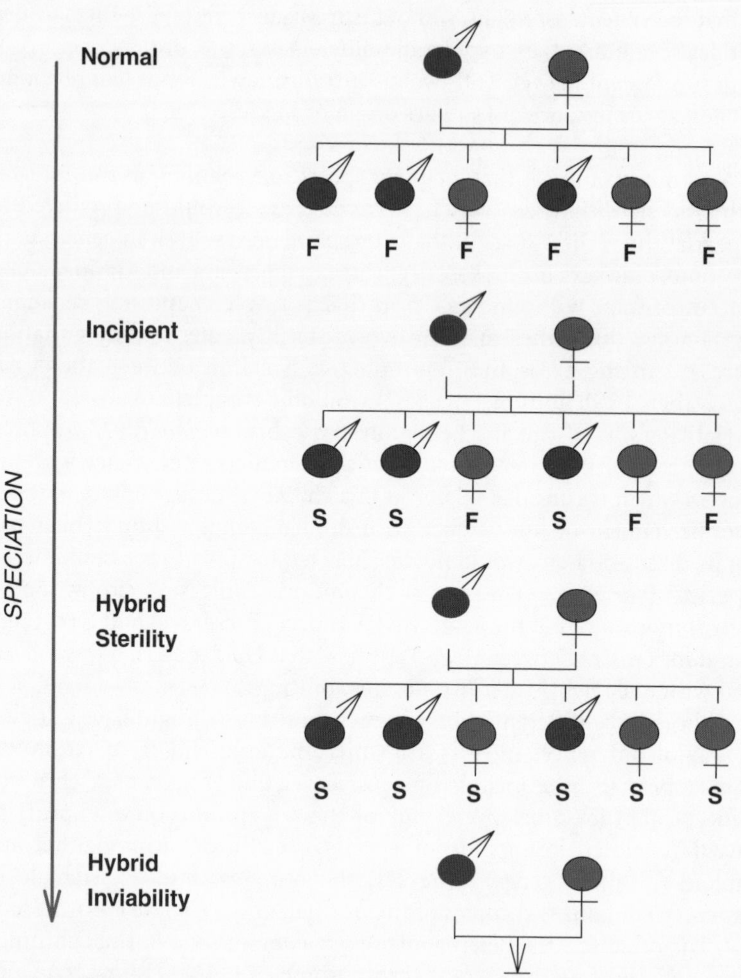

Figure 19.1
Progression towards complete reproductive isolation between two "lines," such that they become "species." In the normal situation (top) there are equal numbers of males ($\male$) and females ($\female$) among the progeny of a cross, and all are fertile (F). Incipient speciation is marked by preferential loss of fertility (sterility; S) in the heterogametic sex (in this case males). Reproductive isolation is complete when this progresses to sterility in all offspring (not necessarily accompanied by decrease in number of offspring). This hybrid sterility barrier may be replaced by hybrid inviability (bottom: initially fewer viable offspring, progressing to no viable offspring). These two post-zygotic barriers might finally be replaced by pre-zygotic barriers (e.g., inability to copulate for anatomical or geographical reasons).

with three important implications: (i) Deleterious recessive mutations are expressed (e.g., haemophilia in human males). Individuals containing these mutations will tend not to reproduce and so fewer deleterious mutations should be passed on to future generations. (ii) It is not possible to repair DNA damage on the basis of information contained in an allelic chromosome. Without accurate repair the load of deleterious mutations passed on to future generations might increase (Bernstein and Bernstein 1991; Gavrilov et al. 1997). Eternally locked in the male generation, the Y chromosome would never have had the opportunity to repair damage on this basis and this may have contributed to its degeneration. (iii) Since gene dosage is often a major determinant of gene product dosage (Figure 18.1), the dosage of X chromosome-encoded gene products (proteins) in males should be half those in females. This discrepancy might either be accepted by the organism ("dosage tolerance" or "dosage imbalance"), or be adjusted ("dosage compensation" or "dosage balance").

Studies by Muller with fruitflies established the existence of a mechanism to adjust the concentrations of X-encoded gene products so that these were equalized between the sexes (Muller 1948). It is now known that this involves an increase in the expression of the single X chromosome in the male fruitfly to equal the combined expression of the two X chromosomes in the female. In mammals equalization is achieved by inactivating one of the X chromosomes in the female (Lyon 1992).

EVOLUTION OF DOSAGE COMPENSATION

Normal individuals are "euploid," meaning that they have the normal number of chromosomes. For humans this means 22 pairs of autosomes (22 of paternal origin and 22 of maternal origin), and either two X chromosomes (females) or a single X and a single Y chromosome (males). "Aneuploidy" means that there is an abnormal number of chromosomes. For example, an individual might have only one copy of an autosome, and so would be "monosomic" for that autosome. This form of aneuploidy is usually lethal. Sometimes the whole chromosome complement is replicated to produce a state known as polyploidy (see chapter 7). Organisms may be triploid (3 sets of chromosomes), or tetraploid (4 sets of chromosomes). Polyploidy is *not* necessarily lethal, implying that the *ratio* of chromosomes and/or their products (not the absolute quantities) must be kept constant.

The question of what drove the evolution of dosage compensation was partially addressed by Ohno (1967 ch.6, 82) who noted that males could be regarded as aneuploid for the X chromosome: "The hemizygous existence for all the genes on the X should be very perilous since monosomy for even the smallest autosome (only one-fourth the size of the X) is

apparently lethal in man." Why aneuploid states are lethal was not discussed.   Studies with the fruitfly led Lindsley and coworkers to conclude
(1972) that: "The deleterious effects of aneuploidy are, in the main,
caused by the *additive* effects of genes that slightly reduce viability and
not by the individual effects of a few aneuploid-lethal genes among a
large array of dosage-insensitive loci."

Orr (1990) when considering why polyploidy is rarer in animals than
in plants, took the issue further. He first noted: "the difficulty in estab-
lishing a tetraploid line in organisms with a genetically degenerate sex
chromosome: although polyploid speciation does not necessarily disrupt
sex determination in such species, it *does* invariably disrupt the balance
of X chromosome [gene products] relative to autosomal gene products
[which is] normally maintained by dosage compensation" (Orr's italics).
He noted further that in mammalian females: "the number of Xs that
remain active increases with the number of autosomal sets" [so that] "the
doses of X-linked and autosomal genes have been *fine-tuned* by natural
selection to ensure *proper interactions* between these loci." The nature of
the "proper interactions" and why they had to be "fine-tuned" were not
discussed.

The importance of the *ratio* of sex chromosomes to autosomes *both*
for dosage compensation and for sex determination is well recognized in
fruitfly and other systems. Since in mammalian systems, ratio-depend-
ence seems to have been retained only for dosage compensation, it is likely to be critical for this process. However, it is difficult to imagine how sex
chromosomes and autosomes would sense each other's dosage directly
and it seems more likely that the quantitation is carried out at the gene-
product (mRNA or protein) level. Furthermore, noting that polyploid
cells are larger than euploid cells (thus presumably keeping the concentration of cytoplasmic components constant), it seems likely that the
quantitation involves some relationship between the concentrations of
the products of sex chromosomes and autosomes (rather than a relationship between their absolute quantities).

THE PARADOX OF "EXCEEDINGLY
MINUTE DIFFERENCES"

Muller (1948) approached the problem of the evolution of dosage compensation in fruitflies by first noting the relationship between the concentration of a protein and the activity of that protein as observed in a
quantifiable phenotype. As the dose of a gene (and hence of the corresponding protein) is increased, there would be an approximately linear
increase in the character assayed. However, this would only hold as long
as the dose of the protein were limiting. Above a certain dose something

else would become limiting and the dose-response curve would tend to plateau (Figure 18.1). Muller noted that the dose of most wild-type gene products (protein) corresponds to a point well along the plateau so that a decrease in dose by half, due to hemizygosity, would still leave sufficient gene product to guarantee maximum activity (i.e., the phenotype would still correspond to a point on the plateau of the dose-response curve). In this circumstance there would have been *no selection pressure*, based on differential gene-product function, for dosage compensation to have evolved. He describes this paradox as follows:

> The effects of individual genes, whether on the X or other chromosomes, are so near their saturation levels as to make direct discrimination between one and two doses *impossible*. Should not the very fact that most of these genes are so near their saturation level make dosage compensation unnecessary? Why should there be a perceptible advantage in going through the motions of equalizing them still further?

Unlike Sherlock Holmes (see chapter 5) he was not prepared to eliminate the "impossible" and seek the "improbable." In his paper boldly entitled "Evidence for the precision of genetic adaptation" Muller concluded:

> The compensation mechanism must be concerned with the equalization of exceedingly minute differences. Dosage compensation has in fact become established because of its advantage in regulating more precisely the grade of characters whose variations in grade, *even without it*, would be exceedingly minute (Muller's italics). ... [Thus] the selective forces that established it must depend on such minute advantages.

One of the architects of "the modern synthesis" in evolution biology of the 1930s (Forsdyke 2001g), Muller here interpreted his results as demonstrating the power of Darwin's natural selection.

FINE-TUNING OF INTRACELLULAR PROTEIN CONCENTRATIONS

A consideration of the fine-tuning of the concentrations of individual proteins within cells provides a not too improbable solution to Muller's paradox (Forsdyke 1994b; 1995a, b). Over a series of generations, a Y chromosome and its descendants exist in a succession of male cells (i.e., the Y chromosome defines a cell as male). This path may be expressed as $M \rightarrow M \rightarrow M \rightarrow M \rightarrow M \rightarrow M \rightarrow M \dots$. Over evolutionary time, factors such as the transcription rates of genes on the Y chromosome and the stabilities of their products (mRNAs and proteins) could have become fine-tuned to the needs of this relatively stable intracellular environment. A relatively constant concentration of each gene product could have become established.

On the other hand, the X chromosomes and the autosomes and their descendants randomly alternate, from generation to generation, between male and female cells. A typical path might be $M \rightarrow F \rightarrow F \rightarrow M \rightarrow F \rightarrow M \rightarrow F....$ Over evolutionary time it would be difficult to fine-tune both for cells containing a second X chromosome and for cells containing a solitary X chromosome. By inactivating one X whenever two are present (the mammalian model), or hyperactivating the X whenever one is present (the fruitfly model), the intracellular environment would be *stabilized* and fine-tuning could occur (Figure 19.2). It should be noted that Y chromosome-encoded proteins are at very low concentrations and may be partly matched in females by a few proteins encoded by uncompensated genes on the X chromosome.

SUSTAINING AGGREGATION PRESSURE

What is it about the intracellular environment that has to be stabilized? Apart from their individual specific functions, proteins have *collective* functions. One well-known collective function, the Donnan equilibrium, affects the distribution of salts between intracellular and extracellular compartments (Hitchcock 1924). Another possible collective function relates to the problem of how a cytotoxic T cell would recognize a cell which had been infected with a virus (Forsdyke 1994b; 1995a). Some virus proteins are partly degraded within the host cell into short amino sequences (peptides). These peptides are then displayed at the cell surface as a complex with a host MHC protein (major histocompatibility complex protein). When presented in this way the foreign ("not-self") peptides are recognized as such by the host's cytotoxic T cells and the virus-containing cell is destroyed. The cell can then be replaced by the mitotic division of uninfected host cells.

The problem with understanding this is that a cell's *own* proteins ("self") are continually being degraded and so should be available for display as peptides at the cell surface. MHC-self peptide complexes are indeed found. However, the virtue of an *obligatory* display of self intracellular peptides is not readily apparent, as it would seem to require the prior deletion or inactivation of *all* T cells specific for MHC-self peptide complexes. This mechanism for self/not-self discrimination is *extracellular* and affects self-reacting cytotoxic T cells at the time of their original formation. The remaining T cells constitute the repertoire available for recognizing MHC complexes with intracellular peptides of foreign origin.

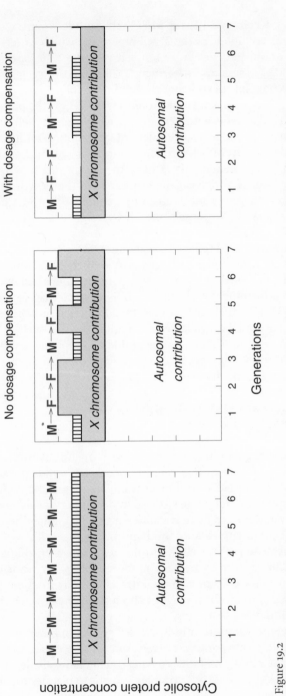

Figure 19.2
Fine-tuning of protein concentration is not possible without X chromosome dosage compensation. Passage of Y or X chromosomes through the generations occurs either in male (M), or female (F) cells. Potential contributions of chromosomes to the cytosolic protein concentration are shown for autosomes, X chromosomes, and Y chromosomes (vertical stripes). The contribution of the Y is much less than shown, so that halving the contribution of the X ("dosage compensation") in female generations would keep the cytosolic protein concentration independent of the sex of the host cell.

Deletion of T cells specific for MHC-self peptide complexes would generate extensive "holes" in the T cell repertoire, which would impair an organism's ability to recognize not-self peptides (Du Pasquier and Blomberg 1982; Vidovik and Matzinger 1988; Schild et al. 1990; Ohno 1991). A mechanism permitting some *intracellular discrimination* between self proteins and not-self proteins would allow the preferential loading of MHC proteins with peptides derived from not-self proteins. This would avoid the logistical problem of competition with a myriad of peptides derived from self proteins (Forsdyke 1994b; 1995a). Thus Hedrick (1992) commented: "It is hard to understand how peptides derived from foreign antigens can compete with the tide of self peptides. ... Perhaps there is a mechanism that could help to sort peptides into those originating from self and those originating from foreign peptides."

We have already hinted (chapter 17) at one way this might be achieved. In the case of an intracellular pathogen (virus), there is present not only foreign protein, but also foreign nucleic acid. Interaction of foreign RNAs with self RNAs (forming double-stranded RNA) would trigger increased expression of MHC proteins *at the same* time as foreign protein degradation products would be available for binding to the MHC proteins (Cristillo et al. 1998, 2001). Ideally, this should be supplemented by some way of recognizing foreign proteins or peptides so that they can preferentially interact with the nascent MHC proteins.

DIFFERENTIAL AGGREGATION OF
FOREIGN PROTEINS

For a cell to distinguish intracellularly between a self protein and a not-self protein (encoded by an intracellular pathogen) would appear a formidable problem. Both classes of protein might, after all, be synthesized on host ribosomes and might be released into similar cytosolic compartments. However, there are three key differences which might be exploited by an appropriate surveillance mechanism.

The first of these is that, relative to not-self protein-encoding genes, each self protein-encoding gene has had more evolutionary time to fine-tune its product concentration to the intracellular environment created by the other self genes with which it has been travelling through the generations (*quantitative* fine-tuning; see chapter 18).

The second, more importantly, is that self genes can generate complex systems for dealing with pathogens. However, pathogens, because of their need to replicate rapidly and disseminate, usually have smaller genomes and thus are less able to encode complex systems to counter the increasingly sophisticated host systems which arise in an escalating arms race. As part of these complex systems, polymorphic self proteins would be selectively advantageous (qualitative fine-tuning; see chapter 15).

The third is that, whereas self protein-encoding genes are essentially at peace with the concentrations at which their products have arrived, the raison d'etre of most foreign pathogens is, at some time within the lifespan of their host, to increase in number. This may imply synthesizing specific proteins at rates resulting in unacceptably high cytosolic concentrations.

If a pathogen with a short generation time and a high mutation rate could gain a foothold, it would readily accommodate to the first difference, but not to the polymorphism associated with the second difference. Furthermore, the third difference combined with the heat-shock response (chapter 18) might be decisive.

How would an unacceptably high cytosolic concentration of a protein be detected? When macromolecules in solution reach a critical concentration it becomes energetically more favourable for them to aggregate like-with-like, than to remain in solution. The aggregation involves a liberation of bound water and an increase in the degree of disorder of the system (entropy). It is a property of disorder (entropy)-driven processes that they are promoted by an increase in temperature, which would be manifest as a fever in warm-blooded organisms (Lauffer 1975; Jarrett and Lansbury 1993; Reich et al. 1995) We should note that, of course, macromolecular aggregation itself is an increase in order, but the *liberation*, by the aggregation, of water molecules that were previously ordered by interactions with the surfaces of the unaggregated molecules, results in a net increase in the total disorder in the system.

The crowded cytosol constitutes an environment which readily drives proteins out of solution when they exceed individual concentration thresholds. This is well recognized from the difficulties encountered when trying experimentally to over-express proteins within foreign cytosols. In such systems, the formation of insoluble aggregates is greatly enhanced by increasing temperature over a physiological range. This leads to the proposal that not-self proteins more readily "trip" the intracellular surveillance alarm because their concentrations are not so fine-tuned to the aggregating properties of the crowded host cytosol as are the concentrations of self proteins (Figure 18.4). In an organism with cytotoxic T cells (or their equivalent) the protein aggregates would be directed to a site for degradation to peptides. Specific peptides would then be displayed in association with MHC proteins at the cell surface.

ROLE OF X CHROMOSOME INACTIVATION

Without dosage compensation there would be a fluctuation in the total intracellular protein concentration within male and female generations

(Figure 19.2). This would imply a fluctuation in the pressure to drive individual proteins out of simple solution when their concentrations exceeded specific concentration thresholds. An increase in activity of the X chromosome in male generations, or inactivation of one X chromosome in female generations, would stabilize the pressure and favour fine-tuning of gene-product concentrations over evolutionary time. A corollary of this is that, in addition to being under evolutionary pressure to preserve *specific function*, genes encoding proteins are also under evolutionary pressures (i) to maintain a *collective function* (namely the pressure to drive individual proteins from solution), and (ii) to maintain individual protein solubilities in the face of that collective pressure.

A disturbance of phenotype resulting from hemizygosity enforced by deterioration of the Y chromosome could involve both the *specific* functions of sex chromosome-encoded proteins and their *collective* functions as proteins per se. While the concentration of a protein in a heterozygote might decrease to a level insufficient to affect specific function (i.e., the concentration of the protein would still correspond to a point on the plateau of the dose-response curve; Figure 18.1), the decrease in concentration of the protein might *itself* be sufficient to affect genetic fitness due to a *substantial*, not "exceedingly minute," influence on some concentration-dependent *collective* protein function, such as the postulated intracellular self/not-self discrimination. This would have driven the evolution of dosage compensation and would appear to resolve Muller's paradox.

WHY SEX?

I have tried to show here that "the biomedical sciences and much of our existence" can best be understood in the light of evolutionary biology. We understand life only to the extent that we fully understand the principles and mechanisms of evolution, and we have here sought a "new synthesis" (White 1978 ch.10, 349) with the correction of genetic errors as an underlying theme.

It was appreciated in the nineteenth century that, mathematically, sexual reproduction appeared disadvantageous relative to asexual reproduction. Winge (1917) suggested that there was a net advantage because sex facilitated the correction of genetic errors (see chapter 8). However, once a new evolutionary path is taken *there is no return*. Once the proto-giraffe "decided" to increase the length of its neck, it needed all four legs for support, so foreclosing the option of developing hands. We have seen in this chapter that once the path of chromosomally-determined sex was taken there were adverse consequences. Perhaps, because it is chemically difficult to confine anti-recombination activity, the full advantages of error-correction were denied to one of the sex chromosomes, which degenerated. The resulting heteromorphism upset the carefully-tuned balance of cytoplasmic gene products, so that there was an adaptive pressure for the evolution of dosage compensation.

The degeneration of one of the sex chromosomes has other consequences which are illuminated by recent genealogical data from European aristocratic families (Gavrilov et al. 1997). If your father conferred on you his feeble Y chromosome, you are a male. As far as we know, your mother, entirely randomly, conferred on you one of her two X chromosomes. These, in turn, she had derived from your maternal grandmother and your maternal grandfather. Does it matter which X you received? Your maternal grandmother's X had coexisted for a generation with another X chromosome (since your maternal grandmother was female) and there was the possibility for error-correction at meiosis. Your maternal grandfather's X had coexisted for a generation with a Y (since your maternal grandfather was male) and there was much less possibility for error-correction at meiosis. However, both maternal grandparents' X chromosomes coexisted for a generation within your mother, so that there was the possibility of correction of any defects in your maternal grandfather's X, on the basis of information contained in your maternal grandmother's X. So the answer is that, if you are a male, it probably does not matter whether you received your maternal grandmother's X or your maternal grandfather's X. On average, your lifespan is likely to be less than females', perhaps because there is less possibility of error-correction of your lone X somatically (Bernstein and Bernstein 1991 ch.12, 255).

If your father conferred to you his X chromosome, you are female. You had no choice but to accept that X chromosome, since your father had only one to confer. Since it was a lone X, there was little possibility of error-correction at meiosis, and you just had to accept what came along "warts and all." On the other hand, your mother conferred upon you randomly one of her two X chromosomes, which, as mentioned above, were present together within her for a generation, providing an opportunity for mutual error-correction. So it probably does not matter which X you received from your mother.

Gavrilov and coworkers (1997) argued that the "warts and all" penalty of the paternal X chromosome you inherited would increase with paternal age. The older your father was when he conferred femaleness upon you by transferring his lone X, the more time there was for errors to have accumulated in his X. Thus, in simple terms, as an XX female you might be either $X_{maternal} X_{paternal-young}$, or $X_{maternal} X_{paternal-old}$. The latter type should be genetically inferior to the former type. A simple and sensitive test of phenotype was used. This was the age at death of the female offspring of young fathers relative to the age at death of female offspring of old fathers. Appropriate controls were carried to rule out maternal age effects. The highly significant answer was that, if your father was old when he passed on his X to join your mother's X and create you, then your age span is likely to be decreased by several years. This simple result underlines much that has been said in this book.

HALDANE REVISITED

Some families have all sons and some all daughters. In most cases, this is due to the random flipping of the "sexual coin." However, some all-daughter families might be a manifestation of Haldane's rule. If the parents were genomically diverged, representing potential incipient species, then their hybrids (the daughters) should manifest hybrid vigour (Figure 8.1). If a son were produced, he should also manifest hybrid vigour, but might be sterile. Among humans, a manifestation of hybrid vigour might be high achievement. Do high-achieving daughters tend to have sisters rather than brothers? Are they usually younger sisters (see above)? If they have a brother, is he a high achiever? It is of note that Haldane, a high achiever in many ways (Forsdyke 2001e), married twice but no children resulted. His only sister, Naomi Mitchison, was both a highly successful author, and highly fertile (Haldane 1961).

SUMMARY

When sex is chromosomally determined, differentiation of sex chromosomes *within* species requires the development of anti-recombination activity which might be analogous to that required for the differentiation of homologous chromosomes *between* species, resulting in hybrid sterility. If so, the sex chromosomes would have already taken a step towards speciation, and hybrid sterility should be detected earliest in the het-erogametic sex. Since this prediction often applies (Haldane's rule), it is possible that the differentiations are mechanistically similar. The degeneration of one chromosome in the heterogametic sex implies that, unless there were some form of compensation, the dose of many gene products would be halved relative to their dose in the homogametic sex. However, the "factor of safety" described in the previous chapter led Muller to deduce that the concentrations of gene products would be so near their saturation levels *with respect to function*, that discrimination between one and two doses would be "impossible." Thus there would have been no selection pressure for the evolution of dosage compensation. This paradox is resolved by considering protein *concentration* rather than protein *function.* The fine tuning of the concentration of an individual protein to the collective pressure exerted by its companion proteins requires that intergenerational fluctuations in that collective pressure be minimized. Dosage compensation prevents the fluctuations and allows fine-tuning, so facilitating the self/not-self discrimination required for detection of intracellular pathogens.

PART FOUR

The Darwinian Struggle for Truth

20   The Philosopher

"Darwinism ... is, indeed, a remarkably simple theory, childishly so, one would have thought, ... But we have good grounds for believing this simplicity is deceptive."

Richard Dawkins (1986 preface xi).

Science conventionally proceeds from hypothesis to experiment, which in turn generates data which support or refute the hypothesis. The deluge of data arising from various genome projects has somewhat altered this sequence. Hypothesis can proceed directly to data, which, in a twinkling, can be analyzed by powerful computer software. Thus in my laboratory in the 1990s we were able rapidly to test some new ideas on evolution. The results were supportive, papers were written, and after minor skirmishes with reviewers, were duly published. Our conclusions seemed so simple – so childishly simple perhaps (see Part 2), that the question arose as to why Darwin and those perspicacious Victorians associated with him had got so confused? Perhaps one of them had not been quite so confused?

SEARCH FOR A VICTORIAN

My search for a Victorian led me to read, for the first time (I shamefully admit), the original *The Origin of Species* (Darwin 1859) and the *Darwiniana Essays* of Huxley (1893). I suspected that the person I was seeking would be one of the immediate "post-Origin" generation, like William Bateson. However, several years after the death of Romanes, Bateson still believed that the "*most serious*" difficulties with Darwin's theory were the small initial variations lacking utility (non-adaptive), and the swamping effect of intercrossing (Bateson 1899, 162). His early researches were concerned with detailed observation of variations, work which was given great impetus by the discovery of Mendel's papers. It was not until 1902 that he began to show a guarded appreciation of the importance of hybrid sterility (see chapter 6).

Indeed, Bateson (1904a, 237) extolled the virtues of the "practical man" who will "stoop to examine Nature" in "the seed bed and the poultry yard." He seemed not to think highly of those (unnamed) with a philosophical bent of mind, who were interested in hybrid sterility achieved by some imaginary form of selection: "For the concrete in evolution we are offered the abstract. Our philosophers debate with great fluency whether between imaginary races sterility grew up by an imaginary Selection ... and for many whose minds are attracted by the abstract problem of interracial sterility there are few who can name for certain ten cases in which it has already been observed."

A disparaging remark by Huxley that Romanes had got it "so hopelessly wrong" (see chapter 3), caused me to ignore his work initially. Provine's 1971 book did not mention Romanes. Eventually, however, I arrived at chapter 7 of Provine's 1986 book on Sewall Wright. Here (ch.7, 208), Romanes was described as "Darwin's protegé," which is, I suppose, near to what these days we would call a graduate student or "post-doc"; if anyone would have been in a position to sort things out, it would have been Romanes. Then (ch.7, 216-17), the magic words "physiological selection" appeared. This was followed by quotations from Romanes. I knew I had found my Victorian.

Romanes (Figure 20.1) was known by his friends as "The Philosopher." He lived in London and his first meeting with Darwin (Figure 20.2) is described in the biography by his wife (Romanes 1896 ch.1, 13-14):

> A letter in Nature attracted Mr. Darwin's notice, and somewhere about 1874 he invited Mr. Romanes to call on him. From that time began an unbroken friendship, marked on the one side by absolute worship, reverence, and affection, on the other by an almost fatherly kindness and a wonderful interest in the younger man's work and in his career. ... Mr. Darwin met him, as he often used to tell, with outstretched hands, and bright smile, and a "How glad I am that you are so young!"

NOT IMPOSSIBLE, YET IMPROBABLE?

My first impression on reading the three volume series *Darwin, and After Darwin* (Romanes 1893, 1895a, 1897), was not just that Romanes had provided a foundation for our studies, but that his account was so clear. This, of course, prompted the question as to why it has had so little impact? This is for historians of science, but I offer some suggestions here in the final part of this book. It seems that, in Sherlock Holmesian style,

Figure 20.1
George Romanes, circa 1878.
Portrait by John Collier.

Romanes had eliminated the "impossible," but "what remained" seemed so "improbable" that few could be convinced that physiological selection "must be the truth." Beyond what he termed a "physiological peculiarity" of the reproductive system, Romanes could not go into specifics as to the cause of the reproductive selection he was postulating. Nor was he able to elaborate upon occasional remarks that speciation (selective fertility)

Figure 20.2
Charles Darwin, circa 1880.

would require the "suitable mating of 'physiological complements'." To some, there may not have appeared much difference between this and the notion of divine creation.

Indeed, the outspoken purpose of people such as Alexis Jordan, whose observations (1873) had been used by Romanes to buttress his case, was to disprove Darwin in favour of divine creation. It may not have added to his scientific credibility that Romanes was the son of a Presbyterian minister, that his major ally was the Reverend John Gulick, and that he had written books such as *A Candid Examination of Theism* (1876), *Mind and Motion* (1885), and *Thoughts on Religion* (1895). In 1879 he declined an invitation from Huxley, well known for his agnostic views, to join the Association of Liberal Thinkers (Huxley 1900 vol.2, ch.1, 3).

ATTACHMENT TO DARWIN

The biography by his wife (Romanes 1896 ch.1, 14) describes Romanes as completely devoted to Darwin and his ideas: "Perhaps no hero-worship was ever more unselfish, more utterly loyal, or more fully rewarded. ... The great master was as much to be admired for his personal character as for his wonderful gifts, and to the youth who never, in the darkest days of utter scepticism, parted with the love for goodness, for beauty of character, this was an overwhelming joy."

At the time of his friendship with Romanes, Darwin was much concerned with "Pangenesis" (Darwin 1875 ch.27, 349). Pangenesis suggested that the gonads were merely the collecting centre for hereditary (and perhaps acquired) information dispersed about the adult body as so-called "generative elements" or "gemmules" (Figure 4.2). In an 1875 letter Darwin wrote (Romanes 1896 ch.1, 39): "I hope with all my heart that you are getting on pretty well with your experiments; I have been led to think a good deal on the subject, and I am convinced of its high importance, though it will take years of hammering before physiologists will admit that the sexual organs only collect the generative elements."

A further letter in 1876 began: "As you are interested in Pangenesis, and will some day, I hope, convert an 'airy nothing' [critical description by German zoologist Haeckel] into a substantial theory ... ." The correspondence implies a sharing, not only of the experimental, but also of the *theoretical*, burden of his life's work. Romanes' experiments to prove the gemmule hypothesis came to nothing, but served to focus his attention on the gonads. Following the death of Darwin in 1882, he devoted much time trying to disprove Weismann's alternative (and correct) proposal that the germ-line (contained in the gonad from the time of embryogenesis) was quite distinct from the rest of the body (soma; Figure 4.2). Only variations in the germ-line could be passed on to future generations. This

Figure 20.3
Alfred Wallace (1823-1913).

had been noted by Hooker in private correspondence (see chapter 2). In 1890 Romanes moved to Oxford and a book eventually appeared (*An Examination of Weismannism*; Romanes 1893), but opinion had already begun to swing strongly in Weismann's favour.

Although a committed experimentalist, Romanes knew, like Darwin, that there already existed abundant information which might bear on evolutionary questions. He was as dedicated to the seeking out and analyzing of existing data, as he was to the generation of further data. However, in a letter to Romanes in 1891, Huxley hinted at the need for more experimentation rather than further theory (Huxley 1900 vol.2, ch.17, 292). In his correspondence with others Huxley was less restrained (see next chapter). Thus, Romanes did not have the support of two most influential figures of their times, Huxley and Bateson. There may have been many like them who were not prepared to take seriously Romanes' three-volume work *Darwin, and After Darwin*.

WALLACE, THE ARCH-DARWINIAN

Romanes and Gulick were also battling against Alfred Wallace (Figure 20.3), who had become even more Darwinian than Darwin. The latter's ideas had undergone repeated modification, but Wallace's were more rigid. Darwin had appealed successfully to Prime Minister Gladstone for Wallace to be awarded a government pension, thus giving him free rein to continue his studies, which extended to supernatural phenomena. Wallace was a leading advocate of the credibility of such phenomena, arguing that natural selection was insufficient to account for the human brain (1869), and publishing *The Scientific Aspect of the Supernatural* (1866), and *Miracles and Modern Spiritualism* (1874).

Wallace was everything that Romanes was not. Wallace was relatively poor, an advocate of spiritualism, a socialist (seeking to nationalize land), an anti-vaccinationist, and, following the death of Darwin, the pre-eminent authority on evolution. At the end of the nineteenth century there was a sense of complacency both in the biological and the physical sciences. It was considered that major problems had been solved, and it was now just a question of sorting out the details (Badash 1972; Silverstein 1989). In this environment authorities carried much weight.

In 1876 the twenty-eight-year-old Romanes and his elder brother, James, were deceived by a "medium" who claimed to be able to communicate with spirits. Romanes wrote two letters to the sceptical Darwin expressing an inclination (short-lived) to believe in the phenomena he had observed. James, who was fourteen when the family left Canada, had a friend in Kingston with an interest in spiritualism and he sent her drafts of the letters. Romanes later (1880) expressed doubt concerning "the ascertained facts of clairvoyance and mesmerism" which had been proclaimed in a letter in *Nature*. This brought a first contact with Wallace. There were two meetings at which Romanes made no mention of his earlier credulity. When Wallace, who was famous as much for his books on spiritualism as for his work on evolution, visited Kingston on a lecture tour in 1887 the lady gave him access to the drafts, from which he made notes (Wallace 1905 ch.36, 309-26).

At that time Wallace and Romanes were engaged in a fierce public debate over the latter's paper "Physiological Selection: An Additional Suggestion on the Origin of Species" (Romanes 1886). Wallace attacked (1886), and Romanes responded (1887). In a book (*Darwinism* 1889 ch.7, 173-9) Wallace presented a modification of his earlier views, aspects of which had been expressed in correspondence (1868) with Darwin (Darwin and Seward 1903 ch.4, 288-99). The correspondence contained a nineteen-point "proof" concerning the role of natural selection in hybrid sterility, which began with point one: "Let there be a species which has varied [note: past tense] into *two forms* each adapted to certain existing conditions better than the parent form, which they soon supplant" [Wallace's italics].

Darwin replied that: "I demur to [the] probability and almost to [the] possibility of ... [point one] as you *start* with two forms [in the same geographical area] which are not mutually sterile, and which yet have supplanted the parent-form."

Two decades later in his 1889 book (ch. 7, 173-9) Wallace admitted: "The preceding argument ... [now decreased to eleven points] depends entirely upon the assumption that *some amount* of infertility characterizes the distinct varieties which are in process of differentiation into species; and it may be objected that of such infertility there is no proof." Under Wallace's scheme, the event which concerned Romanes – the *initiation* of the speciation process – had *already* happened. Wallace dealt with events subsequent to the process of reproductive isolation. The idea that the infertility he noted might relate to what Romanes proposed did not occur to Wallace. In a separate section he described physiological selection as "another form of infertility," which then he proceeded to attack.

Romanes responded publicly again (1890), this time using not only scientific arguments, but also the argument that one should not rely on the judgement of a person of "incapacity and absurdity" whose past judgements were highly questionable, namely, on spiritualism, socialism, and vaccination. Wallace wrote Romanes some sour private letters, which were later published (Wallace 1905 ch.36, 309-26). These letters protested the "appeal to popular scientific prejudice," revealed Wallace's knowledge of Romanes' earlier flirtation with spiritualism as gleaned from the Kingston letters, and threatened to make "known the fact of the existence of these letters and their general tenor." This would show that Romanes' private judgements were not in accord with his public posture of scepticism with respect to spiritualism. In short, Romanes was a hypocrite. A century later there are many reasons to conclude that Romanes' *mature* judgements with respect to spiritualism, socialism, and vaccination were correct. Perhaps his judgement on physiological selection was also correct.

With Romanes' untimely death in 1894, the issues surrounding the physiological selection theory faded from public view. Gulick was without an ally. Provine (1986 ch.7, 218) reports that an important paper of Gulick (1872b) had been transmitted to the Linnean Society by Wallace, but it was only due to the support of Romanes that an 1887 paper was accepted.

Romanes' last and most important work, his posthumous volume III of *Darwin, and After Darwin* (1897), was upstaged by the rediscovery of Mendel's work in 1900. When pondering the question of historical contingency, it is sometimes asked whether people make history, or history makes people? Of course the answer is both, but in the case of Mendel emphasis seems to be on the latter. Mendel could not make his mark until the time was ripe some thirty-five years after his original publication. In the case of Romanes the situation might have been quite different but for his death when only forty-six. His most able contemporary, Bateson, became preoccupied with overcoming the opposition to Mendelism, and appears to have been unaware of the Romanes–Gulick viewpoint (Forsdyke 2001d).

THOMAS MORGAN

Bateson graduated in Zoology from Cambridge in 1883 and then spent two summers working in the U.S.A. with biologist William Brooks. A later student with Brooks was Thomas Morgan (1866-1945), who shared Bateson's view that Darwinian natural selection, as championed by Wallace and Weismann, was insufficient to explain the origin of species. By 1905 Morgan, citing "a number of writers of whom Eimer is perhaps the most important" (Kellogg 1907 ch.10, 281-5), was also veering towards the Romanes-Gulick viewpoint, which he called "the theory of the survival of definite variations:"

> New forms, in the Darwinian theory, are supposed to be created by a process of picking out individual differences... . On the other hand, the theory of the survival of definite variations refers the creation of new forms to another process, namely, to a sudden change in the character of the germ. The creating has already taken place before the question of the survival of the new form comes up. After the new form has appeared, the question of its persistence will depend on whether it can get a foothold... . This distinction appears to me to be not a matter of secondary interest, but one of fundamental importance, for it involves the whole question of the "origin of species." So far as a phrase may sum up the difference, it appears that new species are born; they are not *made* by Darwinian methods, and the theory of natural selection has nothing to do with the *origin* of species, but with the survival of already formed species. Not selection of the fittest individuals, but the survival of the *sufficiently* fit species [all italics are Morgan's].

GULICK'S INFLUENCE

The Romanes–Gulick correspondence indicates that, although acknowledging its relevance, John Gulick had not really appreciated the importance of hybrid sterility. Thus, his son and biographer, the biochemist Addison Gulick (1932 ch.16, 464) notes: "[John] Gulick's writings of 1887 and thereafter give an elaborate discussion of different aspects of sterility, though never quite placing mutual sterility in relief, the way Romanes did, as likely to be the fuse that ignites the whole powder train."

Romanes and Gulick had been separately climbing towards the peak of a high mountain, their heads much of the time lost in the clouds. Every so often the clouds would clear and they would be privileged to views, lonely views, which they could partially communicate to each other, but not to their contemporaries on the slopes below. In 1890 Romanes wrote (Gulick 1932 ch.15, 420): "It appears to be desirable that, as you and I are the only two human beings who recognize the full importance of 'segregation' in all its forms, we should submit to each other our views before publication, in order that we may speak as far as possible with a common voice."

John Gulick was uninfluential despite the contribution of a major monograph (1905). He generated a complex classification of types of sterility, which, combined with his cumbrous prose, must have done more to confuse than to clarify. However, Addison Gulick (1882-1969) continued the crusade. In the 1932 biography of his father he wrote (ch.16, 497-8):

> The problem of physiological isolation has received but little attention during the last two decades, but the modern picture of the mechanism of reproduction and the physico-chemical processes that it involves would have a very large influence on any consideration of the subject today. The immediate cause of infertility between species would be sought at the present time in either chromosomal incompatibilities or maladjustments of a serological nature ... . From several viewpoints the old proposition of Romanes seems today exceedingly plausible, that a rather trivial mutation or group of mutations might set up a barrier of sterility within a species, and cause the two portions to diverge thereafter into well-marked new species. Gulick and Romanes debated the question whether the physiological barrier would precede, accompany, or follow the visible establishment of a varietal form. Today that question would be reworded to ask whether the mutations giving chromosomal ... incompatibility would come early or late in the series of mutations that differentiated the two forms.

THE FORGOTTEN THEORY

Addison Gulick's campaign came too late. By 1932 the high ground of evolution research had been seized by individuals who approached evolutionary questions from a different perspective. The extent to which the Romanes–Gulick contribution had faded from view was evident from Morgan's address at the opening of the 6th International Congress of Genetics:

> The [phenotypic] characteristic showing the greatest effect is the one generally picked out for genetic work. But at the same time there are changes in other organs that are less conspicuous; some of the characters are so little affected or so variable that ... they would give a picture of *continuity*, rather than discontinuity. They would often pass unnoticed were not attention drawn to them by the discovery of the major change. For the

theory of evolution some of these inconspicuous changes may be *more significant* than the more obvious discontinuous changes. In fact, if evolutionary advances are more often through *invisible physiological mutational changes*, rather than morphological ones, we can better understand the paradoxical situation in which taxonomists find themselves, to wit, that the sharp structural differences, that are used for diagnostic separation of species, relate the characters that seem often to be unimportant for the well-being of the individual. The *new point of view* is a *complete reversion* of much of the thinking in which the evolutionary theory indulged in the past.

Of course, apart from the last sentence, Romanes and Gulick would have agreed with much of this. Along similar lines Darlington (1932 ch.16, 480) stressed the pre-eminence of reproductive isolation in some form:

> Species ... are merely the bye-products of evolution depending on a variety of unessential accidents which lead to discontinuity. ... It is clear that discontinuity between species is due to isolation being *followed* by independent variation. Isolation is doubtless often brought about by [geographical] distribution, by change of habit, or by genetic sterility factors; it is also brought about by structural and numerical changes in the chromosomes.

In his popular *Genetics and the Origin of Species* (1937 ch.8, 228-30), Dobzhansky disagreed only on the timing of reproductive isolation: "The maintenance of species as discrete units *demands* their isolation. Species formation without isolation is impossible. ... [However] isolation is necessary but must not come too early." Dobzhansky mentioned John Gulick not at all. He noted that it was Romanes who "originated the oft-quoted maxim 'without isolation or the prevention of interbreeding, organic evolution is in no case possible' ... ," but then sniped: " ... which if taken too literally overshoots the mark."

Addison Gulick attempted to open the argument again in 1938: "With the progress of years we find a striking reinforcement of the scientific cogency of the theory which Romanes and J. T. Gulick championed; namely that a physiological barrier between two otherwise hardly distinguishable stocks may occur frequently, and must have the effect of *initiating* a train of divergent evolution."

SHIFTING BALANCE

Addison Gulick's efforts were to no avail. Judging from Provine's 1971 account, those responsible for what has become known as "*the* modern synthesis" of Darwin's views were quite unaware of physiological selection and spent much time juggling selection coefficients, and the extent and timing of geographical migration, in order to provide a mathematical underpinning for evolution by natural selection. A prominent role was played by Sewall Wright, whose "shifting balance" ideas might have been influenced by Romanes and Gulick. Yet Provine (1986 ch.1, 24) tells us that Wright only read the first two volumes of *Darwin, and After Darwin.* The crucial third volume seems to have been missed. When interviewed late in life (Provine 1986 ch.7, 229) Wright "had a keen recollection of how deeply impressed he had been" with a version of the Romanes–Gulick theory as presented in Kellogg's *"Darwinism Today"* (1907). The latter text, while mentioning the theory as a form of reproductive isolation, does not mention hybrid sterility in the same context. In a 1939 monograph (ch.7, 47-57), Wright cites Gulick (1905) as:

The first to appreciate the possibility of a random drifting apart of isolated races as a statistical consequence of inbreeding, though not, of course in Mendelian terms. ... The conclusion that the accumulation of non-adaptive differences, arising under local inbreeding, may play a major role in the building up of the differences between species and genera, does away with the argument that these categories must have arisen at single steps by the occasional occurrence of major mutations in those cases where selection cannot be invoked.

There is citation of John Gulick, but not of Delboeuf or Romanes, in all except volume 2 of Wright's four-volume work *Evolution and the Genetics of Populations* (1968, 1969, 1977, 1978).

Harvard evolutionist Stephen Gould in a 1980 paper "Is a new and general theory of evolution emerging?" noted a new readiness to entertain "chromosomal alterations as isolating mechanisms" (White 1978; King 1993), and pointed out:

Some of the *new* models ... regard reproductive isolation as potentially primary and non-adaptive rather than secondary and adaptive ... . In ... chromosomal speciation, reproductive isolation comes *first* and cannot be considered an adaptation at all. It is a stochastic [random] event that establishes a species by the technical definition of reproductive isolation. To be sure, the later success of this species in competition may depend on its subsequent acquisition of adaptations; but the origin itself may be *non-adaptive*. We can, in fact, *reverse* the conventional view and argue that speciation, by forming new entities stochastically, provides raw material for selection.

The spirits of Romanes and Gulick should be chuckling! Gould concluded: "I recommend no return to the antiquated and anti-Darwinian view that mysterious 'internal' factors provide direction inherently ... . They channel and constrain Darwinian forces; they do not stand in opposition to them."

Of course, Romanes never held that internal factors "provide direction inherently" (the doctrine of orthogenesis; see Kellogg 1907 ch.10, 274-85). Romanes' internal factors created conditions such that products of the action of Darwinian forces would be preserved. In this way the factors did not stand "in opposition to them." Intriguingly, when distinguishing an "earlier Darwinian orthodoxy" (e.g., Wallace, Weismann) from a modern Darwinian orthodoxy (e.g., Fisher, Muller, Dawkins) Gould (1980) cited the second, but not the critical third, volume of *Darwin, and After Darwin*. Romanes was correctly portrayed as a critic of the earlier orthodoxy with its postulate that all acquired characters are necessarily adaptive.

All three volumes were cited by Gould's Harvard colleague Ernst Mayr in *The Growth of Biological Thought* (1982 ch.12, 564-5). Here one encounters the comment: "Romanes [and] Gulick ... made no clear separation of geographical and reproductive isolation, nor of individual and geographical variation, and often dealt with speciation as if it was the same as natural selection. The confusion is particularly painful in the writing of Romanes, who invented the misleading term 'physiological selection' for reproductive isolation." It is fortunate indeed that I came across Provine (1986) before Mayr (1982)!

"KEEP OUT OF EVOLUTIONARY BIOLOGY"!

Although he had studied mathematics in his first two years at Cambridge (Barnes 1998), Romanes himself had been sceptical concerning "numerical computation involving the doctrine of chances," remarking (1897 appendix B, 158): "in reference to biological problems of the kind now before us, I do not myself attach much importance to a merely mathematical analysis. The conditions which such problems involve are so varied and complex, that it is impossible to be sure about the validity of the data upon which a mathematical analysis is founded."

This was perhaps a harbinger of Waddington's and Mayr's attacks on the "arid calculations of the mathematical population geneticists," which is well documented by Provine (1971 ch.5, 165-6; 1986 ch.13, 477-84). Haldane (1964) even felt the need to defend the quantitative approach in an article entitled: "In defence of bean bag genetics." Huxley in 1864 had eluded to such sophistry (Huxley 1900 vol.1, ch.18, 258) noting "three classes of witnesses, liars, damned liars, and experts." Ronald Fisher, the panselectionist and founder of modern biostatistics, would hardly have approved of Moroney's mutation (1951 ch.1, 2) to "lies, damned lies, and statistics."

In his *Evolutionary Genetics* John Maynard Smith (1989 preface, vii) warned: "To paraphrase Mr. Truman, if you can't stand algebra, keep out of evolutionary biology"! Whether this attitude has assisted or hindered the "unwearied endeavour" to understand the origin of species should be grist-for-the-mill of future historians of science. In a letter to Darwin in 1880, Romanes noted (Romanes 1896 ch.2, 95):

> The mathematicians must be a singularly happy race, seeing that they alone of men are competent to think about the facts of the cosmos. ... Mathematics are ... the sciences of number and measurement, and as such, one is at a loss to perceive why they should be so essentially necessary to enable a man to think fairly and well upon other subjects. But it is, as you once said, that when a man is to be killed by the sword mathematical, he must not have the satisfaction of even knowing how he is killed.

In human societies priestly groups have always tended to seize the high ground. Those who occupy such ground in modern times not only find it easier to continue their dominance, but come to act as "gate-keepers," those peer-reviewers who decide what the rest will be allowed to read, and whose research will be funded. This has its dangers (see next chapter).

A NEW APPROACH

Of course, complete sterility means that there are no progeny. Thus, geneticists are denied their most powerful experimental approach, the crossing of organisms and the observation of progeny. Faced with this impasse, there would seem to be two alternatives (of which Bateson was well aware): either study the exceptions to the rule (from which some progeny can be obtained) in the hope that something useful will turn up, or devise a new approach. Despite his inclination to treasure exceptions, Bateson's intuition (1909a, 227-30) was that: "Not till knowledge of the genetic properties of organisms has attained a far greater completeness can evolutionary speculations have more than a suggestive value. ... The time is not ripe for the discussion of the origin of species." Bateson's viewpoint, however, did not discourage Dobzhansky and his followers of the "genic" school from initiating genetic studies on the exceptions (Coyne 1992; Orr 1996; Coyne and Orr 1998).

At the conclusion of the biography of his father, Addison Gulick (1932 ch.16, 499) wrote:

> We may anticipate that the evolutionary problems which were outlined by Romanes and Gulick will again be reinstated in more perfect form than is today attainable. Until that time arrives it will remain

impossible to estimate the relative importance of the several forces that cause the progressive transformation of species living under the conditions that nature provides.

Two decades were to pass before biochemists Erwin Chargaff (1951) and Gerard Wyatt (1952), adopting a completely new approach, provided critical evidence which, I have suggested in Part 2 of this book, was to lead to the solution of the "more ultimate problem": What was the chemical basis of the "physiological peculiarity" of the reproductive system? What were the "physiological complements" whose "suitable mating" was required for selective fertility of a type which could generally account for the origin of species (Romanes 1897 ch.5, 94)?

SUMMARY

Misled by the disparaging remarks of Romanes' contemporaries, my search for a Victorian was initially fruitless. A clue in Provine's 1986 biography of Sewall Wright set me on the right track. Much in reverence of Darwin, Romanes was late in discarding certain aspects of the "gemmule" hypothesis. This may have been to the detriment of his own scientific credibility. Sadly, opposition by Wallace and others, the discovery of Mendel's work, and Romanes' untimely death, led to the early demise of the Physiological Selection Theory. In the 1930s the son of Romanes' ally Gulick marshalled new evidence in an attempt to rekindle interest, but by then a different breed of scientist, the mathematical population geneticist, had seized the high ground of evolution research.

21 Huxley and The Philosopher's Wife

> "Yes, one day the dearest, the most beloved will be taken from our side, and the death is not the worst that can befall us. There are trials which are harder to bear because they do not come to us straight from God, but from, it may be, the sins of man."
>
> Ethel Romanes (1902)

Overt opposition to Romanes' views was provided by Wallace (see chapter 20). However, the most penetrating was provided by Huxley (Figure 21.1), the Professor of Biology at what is now known as the Imperial College of Science, Technology and Medicine, in South Kensington, London. A man of deep social conscience, towering over most of his Victorian contemporaries, his presence has been felt throughout this book.

Letters from Michael Foster and Huxley smoothed Bateson's passage to field studies in Russia (1886-1887), and Bateson sent Huxley a copy of his data-laden *Materials for the Study of Variation* (1893), where there is a brief reference (ch.16, 425) to Romanes' practical research on jelly fish (1885), *not to his ideas*. Huxley replied (Huxley 1900 vol.2, ch.22, 372): "How glad I am to see ... that we are getting back from the region of speculation into that of fact again. There have been threatenings of late that the field of battle of Evolution was being transferred to Nephelococcygia [nonsense]. I see you are inclined to advocate the possibility of considerable 'saltus' [jump] on the part of Dame Nature in her variations. I always took the same view, much to Mr. Darwin's disgust, and we used often to debate it."

It is likely that Huxley's disparagement of "paper philosophers" who had got it "so hopelessly wrong" may have directed Bateson away from Romanes' writings (as, initially, it did me). I know of no evidence that Bateson and Romanes ever met or corresponded.

Romanes' great respect for Huxley was evident when he invited him to give the second lecture in the annual series he founded at Oxford. The

Figure 21.1
The elderly Huxley.

first Romanes Lecture was given in 1892 by Prime Minister Gladstone, with Huxley agreeing to play stand-in if the "grand old man" (G.O.M.) could not make it. The third in 1894 was by August Weismann. The extent of Huxley's opposition to "The Philosopher" deeply touched the life of Ethel Romanes, who acted as hostess on these occasions.

THE WOUND

Huxley and his wife planned to stay with George and Ethel Romanes (Figure 21.2) at the time of his Romanes lecture on "Evolution and Ethics" (May 1893). Huxley concluded a letter to Ethel Romanes about the domestic arrangements (1st November 1892; Romanes 1896 ch.4, 287): "Would you like me to come in my P.C. suit? [He was a member of the Privy Council.] All ablaze with gold, and costing a sum with which I could buy oh! so many books! Only, if your late experiences should prompt you to instruct your other guests not to contradict me – don't. I rather like it."

After Romanes' death (23rd May 1894), Ethel Romanes wrote to Huxley asking permission to use this letter (which she enclosed) in the biography of her husband. Huxley replied (letter started on 20th September) referring to a previous letter from her, to which he had not replied (Huxley 1900 vol.2, ch.22, 381): "Pray do not suppose that your former letter was other than deeply interesting and touching to me. I had no more than half a mind to reply to it, but hesitated with a man's horror of touching a wound he cannot heal. And then I got a bad bout of

Figure 21.2
Ethel Romanes (1856-1927), circa 1878.

"liver," from which I am just picking up." Before posting two days later he added a postscript:

I fancy very few people will catch the allusion about not contradicting me. But perhaps it would be better to take the opinion of some impartial judge on this point. I do not care the least on my own account, but I see my words might be twisted into meaning that you had told me something about your previous guest [Gladstone], and that I referred to what you had said. Of course you had done nothing of the kind, but as a wary old fox, experienced sufferer from the dodges of the misrepresenter, I feel bound not to let you get into any trouble if I can help it. A regular lady's P.S. this. P.S. – Letter returned herewith.

Ethel Romanes replied (23rd September 1894): "Dear Mr. Huxley, I am afraid your little joke about "not being contradicted" gives your letter half the charm it presumes in my eyes. Only half. It's such a nice contrast to the solemnities of Mr. Gladstone. (I've a great esteem for G.O.M. too.) I thank you for all your kindness. I don't think any touch of yours will make my wound smart more. I hope you are all right again. My love to Mrs. Huxley. Yours very sincerely, E. Romanes." A postscript implies that the 1892 letter had *not* been "returned herewith": "Please at any rate let me have yr. [your] letter back. [undecipherable] had George and I laughing over it."

The letter appears not to have been returned for some time. Huxley died on 29th June 1895. In a letter dated 16th July 1895 Ethel Romanes wrote: "Dear Mrs. Huxley, Many thanks for the l. [letter] The delay made no difference as I was already [undecipherable]. Yours sincerely, E. Romanes. Do give your mother my love when you see her. I do hope she is well."

If the "allusion about not contradicting me" did not refer to Gladstone, who did it refer to? What were the "late experiences" which might have prompted Ethel Romanes to "instruct" her guests? Even more intriguing, what was the "wound"? This we cannot know for certain in the absence of the original of the letter which Huxley found so "deeply interesting and touching." Why was the letter not to be found among the Huxley papers?

With hindsight, it seems likely that "the wound" relates to Huxley's "contradiction" of George Romanes' development of evolutionary theory in the decade following Darwin's death in 1882. Our understanding of this period is incomplete. Biohistorian John Lesch (1975) has observed: "The development of evolutionary theory in the two decades from Darwin's death to the turn of the century remains very largely *terra incognita* for the historian." Evolutionary theory was then much the province of an academic establishment which was either digesting and responding to creationist attacks on Darwin's theory, or engaged in systematically classifying the various species of animals and plants. The latter was an unending task, but a relatively safe academic haven allowing escape from more difficult problems. Thus, the advancement of evolutionary theory was left to those who were either independently wealthy (Darwin and Romanes), or whose other duties left time for such matters (Gulick).

DARWIN'S CIRCLE

The collaboration with Darwin began in 1874 when Romanes was twenty-six. He met and corresponded with Darwin's circle of friends (including Joseph Hooker, Thomas Huxley, Herbert Spencer, and Alfred Wallace), and also Darwin's sons (including Francis who was about his own age). His rise was, as they say, meteoric. In 1875 Darwin, Hooker, and Huxley supported his membership of the Linnean Society, and in 1879 he was elected a Fellow of the Royal Society.

Apart from work on the origin of species, "The Philosopher" wrote poetry and books on theology and psychology. When in 1879 he declined an invitation from Huxley, well known for his agnostic views, to join the Association of Liberal Thinkers, on the grounds that society at large was not yet ready for agnosticism, Huxley fired back (Huxley 1900 vol.2, ch.1, 3): "I quite appreciate your view on the matter, though it is diametrically opposed to my own conviction that the more rapidly truth is spread among mankind the better it will be for them. Only let us be sure that it is the truth."

This is but one small example of Huxley's life-long quest to expunge all forms of mysticism and humbug from a society plagued by them. His caveat on truth is of particular interest in view of his relationship with Romanes. In most cases truth was on Huxley's side, such as his attack in *The Times* (1890-1891; see Huxley 1896 ch.5, 237-334) on what we might now recognize as the fascist path being taken by "General" Booth, founder of the Salvation Army.

DARWIN'S MANTLE

Romanes' biological studies continued after Darwin's death in 1882. In 1885 he reviewed for the scientific journal *Nature* the book *Evolution without Natural Selection: or The Segregation of Species without the Aid of the Darwinian Hypothesis* by Charles Dixon. This book emphasized, as had Gulick, the importance for evolution of non-adaptive variations, which lacked apparent utility. Romanes' review gave no hint of the conceptual explosion to follow. On 5th May 1886 a number of distinguished members of the Linnean Society may have received notes similar to the following to Professor Meldola (Romanes 1886c): "My dear sir, I hope you may find it convenient to attend the next meeting of the Linnean Society, which takes place tomorrow at 8 o'clock. I am to read a paper on a new theory upon the origin of species, and should like to know what you think of it. To me it appears a theory of considerable importance, but on this account I want to expose it to the best criticism. G.J. Romanes."

The paper was entitled "Physiological Selection: An Additional Suggestion on the Origin of Species." It is related (Thiselton-Dyer 1888b) that when presenting this theory of "reproductive isolation," Romanes "began by saying that he regarded it as the most important work of his life." The paper provoked an editorial on 16th August 1886 in *The Times*:

Mr. George Romanes appears to be the biological investigator upon whom in England the mantle of Mr. Darwin has most conspicuously descended. During many years he frequently and exhaustively discussed the whole philosophy of evolution with the distinguished author of "The Origin of Species," and thus he is in the best position for continuing and extending his work, ... He has lately read before the Linnean Society a remarkable paper, ... which, if it be generally accepted, constitutes the most important addition to the theory of evolution since the publication of "The Origin of Species." The position that Mr. Romanes takes up is the result of his perception, shared by many evolutionists, that the theory of natural selection is not really a theory of the origin of species, but rather a theory of the origin and cumulative development of adaptations. Thus, it fails to account for the erection of varieties, which are mutually fertile, into species which are not mutually fertile ... . Thus, many domestic varieties differ from one another much more than many species, or even genera, in the natural state, and the features which distinguish them frequently lack any conceivable utilitarian significance. Moreover, natural selection tends to swamp incipient varieties by the influence of free intercrossing. ... At least it is an intelligible theory, while, until now, modern evolutionists have had practically no theory at all adequate to explain the actual state of things. ... Mr. Romanes, in his combination of Scotch theological and metaphysical tendencies with rigid evolutionary science, ... cannot fail to occupy a distinguished place in the history of evolutionary theories.

Having given the paper, and published versions of it in *Nature* (1886a) and in the *Journal of the Linnean Society* (1886b), the thirty-eight-year old Romanes would then have awaited the response of his academic peers, many of them decades older, and some perhaps believing that "Darwin's mantle" was rightfully theirs. No approbation would have been more eagerly sought than that of Huxley. However, *The Times'* implication that Darwin's theory was one of the *adaptation* of pre-existing species, not a theory of the actual *origin* of species, was easily interpreted as critical of Darwin. As related in chapter 20, a public attack was launched immediately by Wallace. Romanes responded publicly, point by point. The debate was protracted and rancorous, and extended to journals both in England and in the United States.

HUXLEY'S ATTACK

Meanwhile, Huxley was publicly silent, but privately active. Having been too occupied to write a major obituary for *Nature* at the time of Darwin's death (Romanes wrote one instead), Huxley was engaged on a more discursive appreciation of Darwin and his work. On 14th February 1888 he wrote to Michael Foster, under whom Romanes had studied physiology at Cambridge (Huxley 1900 vol.2, ch.12, 191): "I am getting quite sick of all the 'paper philosophers,' as old Galileo called them, who are trying to stand on Darwin's shoulders and look bigger than he, when in point of real knowledge they are not fit to black his shoes."

On 9th March 1888 (ch.12, 192) he wrote to Hooker: "I have been trying to set out the argument on the *Origin of Species*, and reading the book for the nth time for that purpose. It is one of the hardest books to understand thoroughly that I know of, and I suppose that is why even people like Romanes get [it] so hopelessly wrong."

Later that year Huxley's long-awaited obituary notice of Darwin appeared in the *Proceedings of the Royal Society*. Here he began with the early work on shell-fish, and then, invoking the words of the master himself, commenced the attack (Huxley 1888 ch.10, 283): "No one, as Darwin justly observes, has a 'right to examine the question of species who has not minutely described many." Having questioned Romanes's credentials (i.e., he was a physiologist not a naturalist, so had not himself described species), he next challenged either his intellect, or the care with which he had applied that intellect (ch.10, 186-7):

Long occupation with the work has led the present writer to believe that the "Origin of Species" is one of the hardest of books to master; and he is justified in this conviction by observing that although the "Origin" has been close on thirty years before the world, the strangest misconceptions of the essential nature of the theory therein advocated are still put forth by serious writers. Although then, the present occasion is not suitable for any detailed criticism of the theory, or of the objections which have been brought against it, it may not be out of place to endeavour to separate the substance of the theory from its accidents; and to show that a variety not only of hostile comments, but of friendly would-be improvements, lose their *raison d'être* to the careful student.

The attack concluded with bold assertions which leave no doubt as to its target, and its vehemence (ch.10, 288): "Every species which exists, exists by virtue of adaptation, and whatever accounts for that adaptation accounts for the existence of species. To say that Darwin has put forward a theory of the adaptation of species, but not of their origin, is therefore to misunderstand the first principles of the theory."

Huxley's viewpoint never changed. On 7th August 1893 he wrote in the preface (vii) to the *Darwiniana* volume of his collected essays:

So I have reprinted the lectures as they stand, with all their imperfections on their heads. It would seem that many people must have found them useful thirty years ago; and, though the sixties appear now to be reckoned by many of the rising generation as part of the dark ages, I am not without grounds for suspecting that there yet remains a fair sprinkling even of "philosophic thinkers" to whom it may be a profitable, perhaps even a novel, task to descend from the heights of speculation and go over the A B C of the great biological problem as it was set before a body of shrewd artisans at that remote epoch.

Two months after Romanes' death, Huxley wrote in the Preface to his *Evolution and Ethics* (1896, v): "... deploring the untimely death, in the flower of his age, of a friend endeared to me, as to so many others, by his kindly nature; and justly valued by all his colleagues for his powers of investigation and his zeal for the advancement of knowledge."

Huxley's last public address was at the Anniversary Dinner of the Royal Society on 30th November 1894, six months after the death of Romanes and six months before his own death. Ethel Romanes perhaps read the report in *The Times* (Huxley 1900 vol.2, ch.22, 389-90):

> I do not know, I do not think anybody knows, whether the particular view which he [Darwin] held will be hereafter fortified by the experience of the ages which come after us; but of this thing I am perfectly certain, that the present course of things has resulted from the feeling of the smaller men who have followed him that they are incompetent to bend the bow of Ulysses [Darwin], and in consequence many of them are seeking their salvation in mere speculation.

HUXLEY'S BULLDOG

In "that remote epoch" of the 1860s Huxley had acted as "Darwin's bulldog" in defending evolutionary theory against the creationist attack. Huxley remained in communication with many of the leading scientists of his day. Indeed, a small group of them, including Hooker, Spencer, and sometimes Darwin as "guest," had met on a regular basis, as the so-called "x-club." Hooker's son-in-law, the botanist William Thiselton-Dyer acted very much as Huxley's bulldog in defending evolutionary orthodoxy against what was perceived as an attack by Romanes. In his Presidential address at the Bath meeting of the Royal Society, which was reported in *Nature* on 13th September 1888, Thiselton-Dyer (1888a) came close to accusing Romanes of casuistry and self-promotion:

> I observe that many competent persons have, while accepting Mr. Darwin's theory, set themselves to criticize various parts of it. But I must confess I am disposed to share the opinion expressed by Mr. Huxley, that these criticisms really rest on a want of a thorough comprehension. Mr. Romanes has put forward a view which deserves the attention due to the speculations of a man of singular subtlety and dialectic skill. He has startled us with a paradox that Mr. Darwin did not, after all, put forth, as I conceive it was his own impression he did, a theory of the origin of species, but only of adaptations. And in as much as Mr. Romanes is of the opinion that specific differences are not adaptive, while those of genera are, it follows that Mr. Darwin only really accounted for the origin of the latter, while for an explanation of the former we must look to Mr. Romanes himself.

An excessive desire for self-promotion is also imputed by one modern commentator (Schwartz 1985) who states that Romanes "was keenly aware that Darwin could assist his career." Romanes replied point by point in *Nature* on 25th October 1888, concluding:

> I have thus dealt with Mr. Huxley's criticism at some length, because, although it has reference mainly to a matter of logical definition, and in no way touches my own theory of "physiological selection," it appears to me a matter of interest from a dialectical point of view, and also because it does involve certain questions of considerable importance from a biological point of view. Moreover, I object to being accused of misunderstanding the theory of natural selection, merely because some of my critics have not sufficiently considered what appears to them a "paradoxical" way of regarding it.

Thiselton-Dyer continued the attack in the 1st November 1888 issue of *Nature*, noting "an underlying obscurity of ideas by which I find myself as often completely befogged," refusing "to follow Mr. Romanes into all his dialectical subtleties," and repeating Huxley's assertion of incompetence:

> Mr. Romanes is not a practising naturalist. His method is the very inverse of that of Mr. Darwin. We know that the latter for more than twenty years patiently accumulated facts, and then only reluctantly gave his conclusions to the world. Mr. Romanes, on the other hand, frames a theory which looks pretty enough on paper, and then, but not til then, looks about for facts to support it. In my view, one is not called upon to give much attention at present to physiological selection. ... I myself have carefully considered it in connection with a variety of facts, and have arrived at the conclusion that it is not a principle of very much value.

Romanes (*Nature* 29th November 1888b) had no difficulty in replying, noting "the needless asperity" of Thiselton-Dyer's tone, and concluding on a positive note that: "If the strength of a theory may be measured by the weakness of the criticism, then I have good reason to be hopeful for the future of Physiological Selection."

Thiselton-Dyer responded (6th December; 1888c) accusing Romanes of what we would today call jumping on the Darwin bandwagon: "What, however, I view with less patience than his unsustained generalizations, is his persistent attempt to place them on the shoulders of the Darwinian theory. I have reluctantly arrived at the conviction that his only excuse for so doing is that he has fundamentally misunderstood that theory." Again, Romanes (1888c) delivered a careful point-by-point reply.

WALLACE'S BULLDOG

Meanwhile Wallace had acquired his own bulldog. Professor Ray Lankester, of University College, London, when reviewing for *Nature* (10th October 1889) Wallace's new book on *Darwinism* (1889), briefly mentioned a rising young Cambridge biologist named William Bateson, and then praised Wallace's treatment of hybrid sterility: "In his chapter on the infertility of crosses, Mr. Wallace treats at length and with admirable effect a very important subject, as to which he is full of ingenious novel suggestions and apposite facts. His criticism of Mr. Romanes's essay, entitled 'Physiological Selection,' appears to me to be entirely destructive of what was novel in that laborious attack upon Darwin's theory of the origin of species."

There followed a series of open letters between Romanes and Lankester. Bateson seems to have had a great respect for Lankester, noting when applying for a position in Comparative Anatomy at Oxford that his own: "Candidature is not in opposition to that of Professor Lankester" (Bateson 1890, 37).

SOAKED IN DARWINIAN LITERATURE

Remarkably, in view of the vehemence of the public attack, there was a long and largely respectful private correspondence between Romanes and Thiselton-Dyer, which is contained in the biography and elsewhere. On 21st March 1890 Romanes wrote bluntly (Romanes 1896 ch.3, 239): "The result is to satisfy me that your 'intelligent friends' must have had minds which do not belong to the *a priori* order – i.e., are incapable of perceiving other than the most familiar relations [i.e., cannot work from first principles]. Such minds may do admirable work in other directions, but not in that of estimating the value of Darwinian speculations." A few days later (26th March) he added (Turner 1974 ch.6, 153):

> Nor am I really "hard" upon my friends of the "treadmill." I believe they are doing excellent work, as long as they stick to their mill – driving the machinery of scientific progress to better efforts than I can in my less laborious life. But when this life enables me – as it has – to soak myself in Darwinian literature for so many years, I cannot help feeling the arrogance of those more professional naturalists who, with many other occupations and without half the study or thought which I have given to this particular subject, seek to ride rough shod ... with all the four hoofs of dogmatism.

Perhaps mellowed by the knowledge that he was shortly to die, in a later letter to Thiselton-Dyer (26th September 1893) Romanes wrote (Romanes 1896 ch.4, 316-17): "Most fully do I agree with all that you say regarding criticism. And, especially from yourself, I have never met with any but the fairest. Even the spice of it was never bitter, or such as could injure the gustatory nerves of the most thin-skinned of men. I have, indeed, often wondered how you and _____ and _____ can have so persistently misunderstood my ideas, seeing that neither on the Continent nor in America has there been any difficulty in making myself intelligible." Ethel Romanes seems to have omitted the names of two of the protagonists from the biography.

Despite her husband's efforts, Thiselton-Dyer was not won round, noting in correspondence to Wallace in 1893 (Marchant 1916 ch.8, 459): "Romanes is an old acquaintance of mine of many years' standing. Personally, I like him very much; but for his writings I confess I have no great admiration. ... Romanes laments over *me* [Dyer's italics] because he says I wilfully misunderstand his theory. The fact is, poor fellow, that I do not think he understands it himself. If his life had been prolonged I should have done all in my power to have induced him to occupy himself more with observation and less with mere logomachy."

In 1895 Thiselton-Dyer spear-headed a public attack of the "Darwinists" on Bateson, which was to continue for more than a decade (Provine 1971).

CONSIGNED TO OBSCURITY

It seems that, certainly among his English compatriots, no quarter was given, either when Romanes was alive, or after his death. His fellow Canadian expatriate, Grant Allen (also born in Kingston in 1848), in a survey with Huxley in 1888 of "A Half Century of Science," only acknowledged Romanes' contribution to psychology, not evolutionary theory (Huxley and Allen 1888). The biochemist Addison Gulick concluded in 1932 (ch.15, 456; ch.16, 464) that even his father John Gulick, Romanes' strongest supporter, had not fully understood Romanes' arguments.

In many respects Romanes, working from his houses in London (using Burdon Sanderson's laboratory at University College) and Scotland (where he had his own laboratory), had been running rings round the academic establishment. It was he who, in his early work, established the existence of a nervous system in jelly fish (medusae) after "one of the greatest authorities on the group," Huxley, had stated: "The majority have as yet afforded no trace of any such structure" (Huxley 1869a ch.2, 24; Romanes 1885b ch.1, 13). It was he who wrote the major

Nature obituary of Darwin in 1882, and it was he who was proclaimed by *The Times* in 1886 as having acquired "Darwin's mantle" and as one who could not "fail to occupy a distinguished place in the history of evolutionary theories."

Furthermore, it was he who had attracted the attention of two leading statesmen. Lord Rosebery had sponsored a lectureship for Romanes at the University of Edinburgh (1886-1891; Romanes 1896 ch.3, 177). Lord Salisbury had lamented Romanes' untimely death in his Presidential address to the Oxford meeting of the British Association in August 1894. Here, Salisbury had first noted the presence of "the high priests of science" (including Huxley who, to his later regret, had agreed in advance to second the address), and had then pointed to the *lack* of "unanimity in the acceptance of natural selection as the sole or even the main agent of whatever modifications may have led up to the existing forms of life."

To make matters worse, Romanes was popular because of his "loveable nature," and his "eloquent, clear and convincing" lecturing style. Among the founders of the Physiological Society, Romanes was considered "unquestionably the most brilliant," and his unguarded "appreciation of his own work" was not seen as vanity because it was simply a "natural and unconscious part of his character" (Sharpey-Schafer 1972).

It seems the scientific establishment's only recourse was to round up the wagons, with Huxley, an authority more respected even than Wallace, at the centre. A hint of approval from Huxley, and the whole situation might have been quite different. While Wallace's attack on Romanes was frontal, the attack of the urbane and articulate Huxley appears largely indirect and aimed mainly at any grassroot support Romanes might have had. There is much irony in this. Although the question of the origin of species still awaits a complete solution, it seems possible in the light of the work presented in this book that hybrid sterility, "the weak point" Huxley had tackled Darwin with in the 1860s, was solved in its essentials by Romanes in 1886. It is probable that no one in the land was better qualified to understand Romanes' ideas than Huxley. On the Continent, even Weismann (1904 ch.32, 280-339) failed him.

So successful were the efforts of Huxley, Wallace and their supporters, that Romanes himself has been consigned to relative obscurity. However, for Romanes' evolutionary *ideas* the consignment was only temporary. As outlined in chapter 20, they have been periodically (and it seems unknowingly) resurrected by such prominent evolutionists as Cyril Darlington (1932), Michael White (1978), Stephen Gould (1980), and Max King (1993). This "chromosomal" viewpoint, has been criticized by some modern evolutionists who prefer an alternative "genic" viewpoint (Coyne and Orr 1998). My modification of the chromosomal viewpoint meets these criticisms (see Part 2).

A further irony has come to light. As indicated by Francis Darwin in 1886, Darwin's personal papers (now published) revealed that, as early as 1862, Darwin had himself toyed with what Romanes came to call physiological selection. However, unable to sort out the issues at that time, he dismissed it ("will not do"; see chapter 5). Ethel Romanes documents in the biography (1896 ch.1, 53) that in 1877 Darwin sent Romanes some notes containing "my early speculations about intercrossing."

PORT ROYAL

And where did Ethel Romanes stand in all this? Did she share George Romanes' anguish as week by week leading authorities spoke out on the pages of *Nature* against his ideas. Profoundly religious, she was, to this extent, a great respecter of authority. Her biography of Romanes shows a good understanding of evolutionary science, and it seems probable that she had read, and may even have helped tone down, her husband's replies. The biography is an eloquent testimony of her unchanging view of the splendour both of her husband's character, and of his scientific ideas. Here she notes (ch.3, 162): "Mr. Romanes read widely, and observed much, and no one less deserved the charge of writing without observing, or of being a 'paper philosopher.' ... There is a scientific orthodoxy as well as a theological orthodoxy '*plus loyal que le roi*', and by the ultra-Darwinians Mr. Romanes was regarded as being strongly tainted with heresy."

Further evidence on her position may be found in her monumental 1907 treatise on the school of thought associated with the religious and intellectual community of Port Royal, which collapsed under the wrath of Cardinal Richelieu and Louis XIV in seventeenth century France (Romanes 1907). In the preface (vii) she speculates that: "Had this school of thought been permitted to exist in the French Church, it is possible, nay probable, humanly speaking, that the fortunes of the Church of France might have been fairer." Later (ch.10, 211), she discusses the relationship between the great seventeenth century scientists René Descartes (1596-1650) and the younger Blaise Pascal, who died at the age of thirty-nine in 1662:

They had some scientific conversation, and there seems to be a doubt as to whether Pascal did not lay claim to the discovery of the pressure of air on mercury, when it was really due to Descartes. The truth probably is that the idea occurred to both. Everyone will remember that Darwin and Wallace were working side by side on a scientific question, and published papers almost simultaneously and with perfect independence. Certainly, the Pascal family generally and Blaise in particular, always regarded

Descartes with great respect, and with that fraternal feeling which in all ages binds scientific workers together in a brotherhood which is close and very delightful, as those who have, even for a time, shared it can testify. There does, however, seem to have been some jealousy on Descartes' part; that jealousy which is sometimes oddly and sadly manifested, towards the brilliant young, by the distinguished old, in the scientific and literary world.

FLURRY AT OXFORD

There had been a flurry of correspondence between Romanes and Huxley a few weeks before his 1893 Romanes lecture. On the 22nd April Huxley wrote: "There is no allusion to politics in my lecture, nor to any religion except Buddhism, and only to the speculative and ethical side of that. If people apply anything I say about these matters to modern philosophies, except evolutionary speculation, and religions, that is not my affair."

In the *Life and Letters* of his father, Leonard Huxley (1900 vol.2, ch.21, 353-4) implies that these words caused Romanes to write back "in alarm to ask the exact state of the case" for fear of offending the Oxford dons. Huxley replied (26th April): "It seems to me that the best thing I can do is to send you the lecture as it stands, notes and all. But please ... consider it *strictly confidential* between us two (I am not excluding Mrs. Romanes, if she cares to look at the paper)" [Huxley's italics].

Their concerns, which with hindsight were probably more related to Huxley's activities concerning Romanes' evolutionary views than to Oxford politics, were allayed, and Huxley wrote (28th April): "My mind is made easy by such a handsome acquittal from you and the Lady Abbess, your coadjutor in the Holy Office. My wife, who is my inquisitor and confessor in ordinary, has gone over the lecture twice, without scenting a heresy. ... I was most anxious for giving no handle to anyone who might like to say I had used the lecture for purposes of attack."

Lively as the mind behind these letters appears, it seems that the Huxley of the 1880s was not the Huxley of the 1860s. Although he may not have fully understood what Romanes was saying, he was quite capable of recognizing his accomplishments. On the basis of a distinguished track record in neurophysiology (Ralston 1944; French 1970), and his long apprenticeship with Darwin, Romanes might at least have been given the benefit of the doubt. Even in modern times the *Science Citation Index* documents regular citation in the 1990s of Romanes' 1881 classic *Animal Intelligence*. The words of Archbishop Piccolomini to Galileo in 1633 seem particularly appropriate (Sobel 2000 ch.26, 300): "You deserve this and worse, for you have been disarming by steps those who have control of the sciences, and they have nothing left but to run back to holy ground."

While defending Darwinian "holy ground," Huxley may have believed that, although Darwin had shown that variation and inheritance could be dealt with in general terms without knowledge of their fundamental physiology and biochemistry, opening these "black boxes" was the most pressing item on the research agenda. Perhaps he wanted the next generation of scientist to be a reductionist generation which would solve these mysteries in its "seed beds" and laboratories. The twentieth century was for those who would emulate William Bateson (as perceived by Huxley), not Romanes. In many respects the reductionist approach may have been too successful. Biologists may now have to relearn what the physicists have long known, that theoretical science and experimental science go hand in hand.

JBS

There was another "flurry" at Oxford in 1892, which may have come to Romanes' attention (Romanes 1918 ch.3, 56). It took the form of a hybrid between the Haldanes and the Trotters. The baby was named John Burdon Sanderson Haldane ("JBS"). On coming down from Cambridge in 1874, Romanes had worked in London, with JBS's great uncle, Burdon Sanderson, who was "next to Mr. Darwin, ... the scientific friend George Romanes most valued and loved" (Romanes 1896 ch.1, 13). Burdon Sanderson was appointed Professor of Physiology at Oxford in 1883, and this paved the way for Romanes' move there in 1890. JBS's father, John Scott Haldane, had been appointed in 1887. With common roots in Scotland, the Romanes and Haldanes often met socially. Mrs. Haldane (1961 ch.15, 158) relates:

My first dinner party was at the Romanes', and he was quite definitely Romanes of Nigg [place in Scotland], not an unfamiliar Oxford professor. I saw a good deal of him during his last illness. The excuse was that he liked to hear about some experimental work of John's and we talked at large about people and things, and on the last few visits rather tentatively about religion. He spoke very restrainedly of the many visits from the clergy his wife believed in. "They mean well," he said, but they tired him terribly, and would try to make him answer questions. On my last visit, he held on to my hands and said, "Don't go, don't go," but what can one do when a man's devoted wife almost pushes one out of the room? I am quite confident that his death-bed confession [of faith in God] was merely the result of being too exhausted to argue. It was not meant as acquiescence to Anglican doctrines.

Figure 21.3
J.B.S. Haldane (1892-1964), circa 1916.
Drawing by M. A. Egerton.

Encouraged to develop an interest in genetics at an early age, it seems likely that JBS was aware of Romanes' work (Forsdyke 2001e). In 1908 JBS and his younger sister Naomi began breeding guinea pigs and found certain characters were inherited together (linkage groups). After the First World War JBS (Figure 21.3) came directly under Bateson's influence noting "his great generosity in helping me with his immense knowledge of the by-ways of the entomological literature" (Haldane 1957). However, the allure of the genic viewpoint was too great and JBS became a supporter, albeit hesitant, of "*the* modern synthesis," with its emphasis on genes and the power of natural selection. JBS was probably the last great disparager of Romanes. When commenting on Huxley's problem (see chapter 3) he confidently stated (Haldane 1929): "The stumbling block in the past has been the failure to find, between varieties, the physiological barrier which often prevents the effective crossing of species. This failure was regarded as a serious but not fatal objection to Darwinism by such men as Huxley and Romanes. It has now been completely overcome."

In an article entitled "The theory of evolution before and after Bateson," JBS, after reiterating Huxley's problem, set out to explain Bateson's unhappiness with the genic viewpoint. Few readers then and since would have recognized that he might also have been firing a last salvo at Romanes (Haldane 1957): "Thus Mendelism appeared to have no immediate bearing on the problem of evolution, except to show that the explanations given *sixty years ago* of how evolution had occurred were almost certainly false."

What happened "sixty years ago"? The year 1897 was the date of the posthumous publication of Romanes' masterpiece, the third volume of his *Darwin, and After Darwin*. The first edition of this and other works of Romanes may be found in JBS's library, which is preserved at the Centre for Cellular and Molecular Biology in Hyderabad.

FATE?

The most golden idea will be of little value if it cannot be marketed. As a case study, the relationship between Huxley and "The Philosopher" is important because a failure of marketing can be attributed retrospectively to a scientific establishment whose activities, unlike those of most modern scientific establishments, are documented and preserved in the public domain. There may be no great difference in the dynamics of resistance to novel ideas between the Victorian era, and our present era. Then as now, success at research depended on seven major factors: talent, enthusiasm, funding, right research supervisor, a tractable problem, luck, and receptive peers. Romanes had the first five of these, and much of the sixth. Unfortunately at the age of forty-six his luck ran out. When Bateson later elaborated similar ideas his peers also did not understand (see chapter 22). Perhaps Romanes would have understood. Just as Blaise Pascal and others associated with Port Royal might have transformed the fortunes of the Church of France in the seventeenth century, so might the combination of George Romanes at Oxford and William Bateson at Cambridge have transformed the science of the twentieth century. As it is, we are only just beginning to realize what they were trying to tell us.

SUMMARY

Recent developments in evolutionary theory have cast new light on the clash of evolutionary ideas in the two decades following Darwin's death in 1882. The problem of hybrid sterility, "the weak point" which Huxley had himself repeatedly emphasized to Darwin, may have been solved, in its essentials, by Darwin's research associate George Romanes in 1886. However, Huxley and Wallace, with the help of their "bulldogs," gave no quarter to the "The Philosopher." Following his untimely death in 1894, his wife and biographer, Ethel Romanes, continued to believe that his view would ultimately prevail. The dynamics of peer-resistance to novel ideas in the Victorian era may be similar to those of the present era.

22 "We Commend This State of Mind"

"It is doubt that so loves truth that it neither dares to rest, nor extinguishes itself by unjustified belief; and we commend this state of mind to students of species, with respect to Mr. Darwin's or any other hypothesis as to their origin."

Thomas Huxley (1859 ch.1, 20)

Conventional wisdom has it that scientists first review the literature pertinent to their research interest, construct a hypothesis, and then either perform experiments, or examine the ideas and experiments of others that may bear on the hypothesis. However, given time constraints and the volume of the literature, they have to be selective in their review, taking guidance from those who have gone before. If this guidance is inadequate then the literature review may be inadequate, as may be the work that follows.

It is shown here that Romanes' work was not appreciated by his contemporaries, and was overlooked by most who took up the Darwinian torch in the early decades of the twentieth century. The generation that followed relied either on Darwin's work itself, or the summaries of the earlier literature provided by researchers such as Sewall Wright and Theodosius Dobzhansky. That generation, in turn, provided the literature read by most modern researchers. Reverse the process and follow citations in the modern literature back to their sources, and then follow the citations in those sources back to their sources. The paths back to Romanes are difficult to find (e.g., French 1970; Lesch 1975; Gould 1980; Littlejohn 1981; Mayr 1982; Bowler 1983; Schwartz 1985; Provine 1986; Barnes 1998). Romanes may be as guilty as the rest. He cited Mendel in 1881, probably without first checking the article (Olby 1985, 228-30)!

UNWARRANTED OPTIMISM

Twentieth century scientists also overlooked Bateson's contribution. Albeit late in his life, he came to realize the overwhelming importance of hybrid sterility, and carried the issue as far as anyone could, given the primitive understanding of the chemistry of heredity at that time (see chapters 6 and 13). For some the overlooking was a passive process, but for others there was a quite ferocious opposition. A hint at the forces at work may be derived from an apparently innocent remark at a conference on progress in AIDS research. The director of the U.S. National Institute of Allergy and Infectious Disease declared that "pessimism is destructive" (Cohen 1997). But unwarranted optimism can be equally destructive. An unwillingness to face the difficulties and uncertainties in an area where there are still enormous gaps in our knowledge will almost certainly slow progress. Until knowledge is more complete, strange facts and ideas not easily accommodated by current dogma nevertheless deserve attention. In this circumstance, pessimism regarding the conventional wisdom can be highly constructive. Indeed, Romanes' mentor Burdon Sanderson noted in 1875 that "so long as uncertainty exists, there is nothing to be so much avoided as that sort of clearness which consists in concealing difficulties and overlooking ambiguities" (Romano 1993, 176).

Already deeply engaged in breeding experiments when Mendel's work was rediscovered at the end of the nineteenth century, Bateson immediately perceived its significance. In the words of his wife, Beatrice Bateson (1928, 70): "He was over the stepping stones and away, scrambling up the further bank whilst the Biometricians, chiding, were still negotiating the difficulties of the first step." However, because "unit factors" (genes) explained *so much*, it was tempting to assume that they explained *everything*. As related here, by virtue of his high expertise in the area, Bateson was not so easily satisfied. Meanwhile, in possession of an incomplete but experimentally productive "genic" hypothesis, Morgan and his school of fruitfly researchers, which included Muller (Forsdyke 2001g), were "over the stepping stones and away." This time Bateson appeared to be "negotiating the difficulties."

While Romanes' contributions to evolutionary theory are merely forgotten, Bateson has been "bit by him that comes behind" over the decades since his death. Like Romanes, he was concerned with truth, not just politically acceptable truth. However, a different breed of scientist was coming to dominate scientific opinion. Thomas Huxley had not been supportive of Romanes (see chapter 21), but he would probably have been even less supportive of a complacent grandson (Julian), who optimistically described the consensus reached in the 1930s as "*the* modern synthesis" (Huxley, 1942).

The arch-selectionist Fisher (1930 ch.6, 124) described Bateson's *Materials for the Study of Variation* (1894) as "a work, which owed its influence to the acuteness less of its reasoning than its sarcasm." In a major study "Bateson and chromosomes: conservative thought in science," historian William Coleman (1970) was unable to see that in questioning the role of chromosomes, Bateson was struggling to grasp their informational essence (see chapter 6). The systematist Ernst Mayr (1980 ch.1, 6) classified Bateson as among those geneticists who failed to understand evolution, implying that others understood it better. The plant geneticist Ledyard Stebbins (1980 ch.5, 146) in an article "Botany and the synthetic theory of evolution" blamed Bateson together with Morgan for "delaying *the* synthesis." Regarding the position he perceived to have been taken by the "mutationists of the early part of this century" Dawkins (1983 ch.20, 419) commented: "For historians there remains the baffling enigma of how such distinguished biologists as De Vries, W. Bateson, and T. H. Morgan could rest satisfied with such a crassly inadequate theory. ... The irony with which we must now read W. Bateson's dismissal of Darwin is almost painful."

To the question "Is a new evolutionary synthesis necessary?" (Stebbins and Ayala 1981), the answer was generally negative. In 1996 a paper by H. Allen Orr entitled "Dobzhansky, Bateson, and the genetics of speciation" appeared in the "Perspectives" section of *Genetics*, the highly influential official journal of the Genetics Society of America. Capturing the wide spirit of complacency, with reference to Dobzhansky's *Genetics and the Origin of Species* (1937), the paper declared (here I use ellipses with some licence):

> Sixty years ago, Theodosius Dobzhansky took a large step towards solving "the Species Problem." [In so doing, he] ... put several popular theories of speciation to the sword ... [so effectively] ... that their existence is now forgotten. ... [Dobzhansky] dropped the bomb ... [in his] ... seminal work ... [which] was far more convincing than any preceding it. ... [Thus, Dobzhansky arrived at the] right conclusion, ... [and] resolved a paradox that had stared down evolutionists for half a century. ... [Claims that] something unique, some novel process or novel kind of factor, causes speciation ... [are a] reluctance to render unto Mendel what (it turns out) was clearly Mendel's. ... Bateson was irredeemably confused about evolution, ... [and] was, to some extent, one of the enemies battled against during the modern synthesis. ... As the years wore on, he grew increasingly obsessed and depressed with Darwinism's failure to crack the Species Problem which, to Bateson, meant the origin of hybrid sterility.

Consistent with the notion that "we scientists are notoriously bad historians" (Orr 1999), the paper actually portrayed Bateson as the first *advocate* of Dobzhansky's genic view, a point uncritically relayed by Berlocher (1998 ch.1, 6) in the multi-authored text *Endless Forms: Species and Speciation*. The latter seems likely to become required reading for those entering the field in the first decades of the twenty-first century.

THE "EVOLUTIONARY CONSTRICTION"

When scientific discourse degenerates from the substantive to the personal it is right to be suspicious. Against the confident "hardening of the modern synthesis" protested by Gould (1983), the words of a few lone historians were of little avail. Provine noted (1992, 169): "As a historian I am immediately suspicious when anyone describes his or her views as the 'new' or 'modern' way of seeing things, to be sharply distinguished from the 'old' inferior ways. ... [Indeed] the modern synthesis is scarcely a synthesis at all and should be renamed the evolutionary constriction."

There was a "constriction" since the number of variables considered of major importance was considerably decreased (Provine 1992, 176-7): "The theoretical population geneticists Fisher, Haldane, and Wright argued that evolution in nature could be modelled quantitatively. They demonstrated clearly with their models that evolution within a population could be accounted for by quantitative relationships between *relatively few variables*." By appropriate choices of "selection coefficients" and other fudge factors, the mathematical population geneticists (as the modern biometricians have become known), achieved a quick-fix which led Michael White to lament in his *Modes of Speciation* (1978 ch.10, 349): "... how much still needs to be discovered before we can confidently construct the 'new synthesis' of evolutionary theory some 25 or 30 years from now."

If this were true in 1978, how much truer must have been Bateson's important 1909 article "Heredity and variation in modern lights" in which he emphasized how ignorant we were. In the article he sketched out the general characteristics of the species phenomenon, and pointed to possible solutions either in terms of genes (the approach to be promulgated by Dobzhansky and Wright), or of indefinite "factors" and "substances" the understanding of which awaited a better comprehension of the chemistry of heredity (see chapter 6; Forsdyke 2001d).

BREAKFAST AT ITHACA

Provine (1986 ch.13, 499) predicted that Sewall Wright would be seen as "perhaps the single most influential evolutionary theorist of this century."

After the Toronto meeting of the American Association for the Advancement of Science (see chapter 7), Bateson and Wright took the train east to continue discussions with biologists at Cornell University. In view of his quest for "minds of first-rate analytical power" (see chapter 9), it is likely that Bateson had high hopes for young numerate biologists such as Wright and Haldane. In the biography of Wright, Provine (1986 ch.4, 104) describes the breakfast at an Ithaca hotel, where (presumably as related by Wright) Bateson "was by now fed up with Wright's quantitative analysis and bitterly exploded, condemning the use of biometrician-like statistics in genetics."

Perhaps influenced by Mayr (1982 ch.17, 733), who regarded Bateson as "pugnacious to the point of rudeness," Provine describes this "outburst," as "so typical of Bateson." Yet Bateson's close colleague Punnett wrote (1950): "No one more delightful to work with than Bateson could be imagined. Dominant personality as he was, he was never domineering. Only once can I recall his having lost his temper with me, and then he had every justification because I had imported a trio of Silky fowls without his knowledge."

Haldane noted in 1957 that "his anger was largely reserved for inaccuracy and loose thinking, and for certain types of injustice," and that "he showed no signs whatever of a senile failure of original thought." Can we suppose that Bateson employed anger, like a tool, as a last resort in an attempt to shake some sense into Wright? In vein, the Victorian-Edwardian may have been trying to turn from his preoccupation with genes one who, he may have anticipated, would become "the most influential evolutionary theorist" of the twentieth century. What more could Bateson have done?

ADVOCACY

Although his final acts of advocacy were to fail, much of Bateson's professional life was spent in advocacy with some success. Many textbooks state simply that Mendel's work went unnoticed until the turn of the century, when the new science of genetics began. However, the entrenched scientific establishment which had disregarded Romanes showed little inclination either to explore the implications of Mendel's work, or to support those who wished to do so. Opposition by his old friend Weldon and by Thiselton-Dyer is well documented by Provine (1971 ch.3, 62-87). Punnett (1950) recollected that: "In spite of the success of the [1904] Cambridge meeting in getting Mendelism a hearing, the older generation of biologists endorsed Weldon's hostility and the pens of Alfred Russel Wallace, Professor Poulton, and Professor J. Arthur Thomson were soon engaged in attempting its belittlement. In this they were supported by [the journal] *Nature* ... ."

These scientists would have acted as reviewers for scientific journals such as *Nature*, and thus: "It was a difficult time for struggling geneticists when the leading journals refused to publish their contributions, and we had to get along as best we could with the more friendly aid of the Cambridge Philosophical Society and the Reports to the Evolution Committee of the Royal Society." Bateson had written (1889, 161) in "Hybridization and cross-breeding as a method of investigation":

It is perhaps simpler to follow the beaten track of classification or of comparative anatomy, or to make for the hundredth time collections of the plants and animals belonging to certain orders .... If the work which is now being put into these occupations were devoted to the careful carrying out and recording of experiments of the kind we are contemplating [i.e., genetics], the result ... would in some five-and-twenty years make a revolution in our ideas of species, inheritance, variation, and the other phenomena which go to make up the science of Natural History.

He later observed (1904a, 234): "The successes of descriptive zoology are so palpable and so attractive, that ... these which are the means of progress have been mistaken for the end. ... We may hope that our science will return to its proper task, the detection of the fundamental nature of living things." In similar vein Darwin in 1872, when praising a book on natural history, had declared (Darwin and Seward 1903 ch.5, 337): "How incomparably more valuable are such researches than the mere description of a thousand species." This might well be paraphrased today with respect to "a thousand genes" or their products.

Until the importance of Mendel's work was widely appreciated, Bateson felt quite isolated. Using speciation as a metaphor he addressed the Zoological Section of the British Association at Cambridge (1904):

Those whose pursuits have led them far from their companions cannot be exempt from that differentiation which is the fate of isolated groups. The stock of common knowledge and common ideas grows smaller till the difficulty of inter-communication becomes extreme. Not only has our point of view changed, but our materials are unfamiliar, our methods of enquiry new, and even the results attained accord little with the common expectation of the day.

The failure of succeeding generations to come to grips with the species problem, partly reflects the increasing need to specialize in order to satisfy "the common expectation of the day," namely to gain the peer approval so necessary for professional advancement (Forsdyke 1993; 1995g; 2000a). If "truth" is that which is perceived by one's scientific peers, then one should just read what they read and avoid distractions. An inadequate literature review with respect to the sort of "truth" which so concerned Bateson and Romanes carries few penalties, and there are no rewards for those who might be inclined to comment on the emperor's attire. There were few grants in Victorian times and Romanes, like Darwin, was independently wealthy so could finance his own research. However, for many scientists today the choice between hunting with the pack, or not hunting at all, is a very real one. Thus, the process of academic speciation is alive and well. Intellectual isolation by "interdisciplinary oceans of ignorance" ensures that those in "islands of near-conformity" will thrive (Ziman 1996).

SUMMARY

When knowledge in a field is incomplete, strange facts and ideas not readily accommodated by current dogma deserve attention. In this circumstance, pessimism regarding the conventional wisdom can be highly constructive, while the complacency of those who unquestioningly accept that wisdom can be highly destructive. If the thesis presented here has any validity, then Romanes and Bateson were not just *years* ahead of their contemporaries, they were *light years* ahead. Attempts by adherents of "*the* modern synthesis" to portray them as feeble-minded obstructionists may have delayed a true synthesis of evolutionary theory.

Epilogue

In the Prologue I record my surprise at finding that Romanes was born in Kingston, Canada. To appreciate a somewhat surreal aspect of this, the reader should first know a little of my personal history. I was born in London, England, and in 1961 graduated from St. Mary's Hospital Medical School (now part of "Huxley's college," the Imperial College of Science, Technology and Medicine, at London University). After internships in medicine, psychiatry and surgery, I obtained a Ph.D. in biochemistry at the University of Cambridge, where I resided in Christ's College (Darwin's college). Through the employment section of various scientific journals, I then looked for a university position. It so happened that there was an attractive opening at the Department of Biochemistry at Queen's University, Kingston, where I have worked since 1968.

There was a quite random element to this choice. I might have ended up at another university in North America, or in Australia or New Zealand. I might have remained in the United Kingdom. Wherever I settled, it is very likely that I would have continued my Cambridge interests in Theoretical Biology, Molecular Biology and Cellular Immunology, and that in the 1980s I would have turned to Bioinformatics/Genomics (computer analysis of DNA sequences). In the early 1990s I arrived at a relatively simple view of the origin of species (published in the *Journal of Theoretical Biology* in 1996). As related here, this led me to wonder if one of the Victorians close to Darwin might have anticipated me. After following several false trails in my search for a Victorian, I came across Romanes in the fall of 1997 and learned of his origin, not in the United

Kingdom, but in the "colonies," and in the particular colonial city where I lived!

Remarkable enough, but there is more. It so happens that, with my family, I had moved in 1982 to a grey stone house which forms part of a small block in William Street, Kingston. Shortly *after* the move we found that in the 1840s the block had housed Queen's College (see *Queen's: The First Hundred and Fifty Years*, pp. 28-29. Hedgehog Productions Inc., Newburgh, 1990). Further enquiries in 1997 led to the discovery that Professor George Romanes (1806-1871) and his family occupied part of the block from 1846 to 1850 (M. Angus 1986; "Queen's College on William Street," *Historic Kingston* 34, 86-98). The Presbyterian minister had migrated from Scotland in 1834, and had married (1835) another Scottish immigrant, Isabella Gair Smith (1811-1883). James was born in 1836, a second son died while young, and Georgina was born in 1842. George Romanes became Professor of Classics at Queen's College (founded 1841) and a member of the University Senate. 1848 was a particularly good year. He inherited "a considerable fortune" leading to the decision to return to Europe in 1850, and George John Romanes was born (Ringereide 1979). Thus, it seems likely that the latter spent his first two years but a few yards from where these words are being typed!

References

Allen, G.A. (1978) *Thomas Hunt Morgan. The Man and his Science.* Princeton University Press, Princeton.

Ayala, F.J. and Fitch, W.M. (1997) Genetics and the origin of species: an introduction. Proc. Natl. Acad. Sci. USA 94, 7691-97.

Badash, L. (1972) The complacency of nineteenth century science. Isis 63, 48-58.

Baker, S.M., Plug, A.W., Prolla, T.A., Bronner, C.E., Harris, A.C., Yao, X., Christie, D.M., Monell, C., Arnheim, N., Bradley, A., Ashley, T. and Liskay, R.M. (1996) Involvement of mouse Mlh1 in DNA mismatch repair and meiotic crossing over. Nature Genetics 13, 336-42.

Ball, L.A. (1973) Secondary structure and coding potential of the coat protein gene of bacteriophage MS2. Nature New Biol. 242, 44-45.

Barbujani, G. (1997) DNA variation and language affinities. Am. J. Hum. Genet. 61, 101-14.

Barnes, E. (1998) The early career of George John Romanes, 1867-1878. Unpublished B.A. Thesis. University of Cambridge.

Baron, L.S., Gemski, P., Johnson, E.M. and Wohlhieter, J.A. (1968) Intergeneric bacterial matings. Bacteriol. Rev. 32, 362-9.

Barrette, I.H., McKenna, S., Taylor, D.R., and Forsdyke, D.R. (2001) Introns resolve the conflict between base order-dependent stem-loop potential and the encoding of RNA or protein. Further evidence from overlapping genes. Gene (in press).

Bateson, W. (1890) Letter to the Electors to the Linacre Professorship. In *William Bateson, F.R.S. Naturalist. His Essays and Addresses*, pp. 30-37. Edited by B. Bateson. Cambridge University Press, Cambridge, 1928.

– (1894) *Materials for the Study of Variation.* Macmillan & Co., London.

- (1899) Hybridization and cross-breeding as a method of scientific investigation. In *William Bateson, F.R.S. Naturalist. His Essays and Addresses,* pp. 161-71. Edited by B. Bateson. Cambridge University Press, Cambridge, 1928.
- (1904a) Presidential address to the Zoological Section, British Association. In *William Bateson, F.R.S. Naturalist. His Essays and Addresses,* pp. 233-259. Edited by B. Bateson. Cambridge University Press, Cambridge, 1928.
- (1904b) Memoire. In *William Bateson, F.R.S. Naturalist. His Essays and Addresses,* p. 94. Edited by B. Bateson. Cambridge University Press, Cambridge, 1928.
- (1907) The progress in genetics since the rediscovery of Mendel's papers. Progressus rei Botanicae 1, 368-418.
- (1908) The methods and scope of genetics. In *William Bateson, F.R.S. Naturalist. His Essays and Addresses,* pp. 317-33. Edited by B. Bateson. Cambridge University Press, Cambridge, 1928.
- (1909a) Heredity and variation in modern lights. In *William Bateson, F.R.S. Naturalist. His Essays and Addresses,* pp. 215-32. Edited by B. Bateson. Cambridge University Press, Cambridge, 1928.
- (1909b) *Mendel's Principles of Heredity.* Cambridge University Press.
- (1913) *Problems of Genetics,* pp. 238-42. Yale University Press, New Haven.
- (1914) Presidential address to the British Association, Australia. In *William Bateson, F.R.S. Naturalist. His Essays and Addresses,* pp. 276-316. Edited by B. Bateson. Cambridge University Press, Cambridge, 1928.
- (1917) Gamete and zygote. A lay discourse. In *William Bateson, F.R.S. Naturalist. His Essays and Addresses,* pp. 201-14. Edited by B. Bateson. Cambridge University Press, Cambridge, 1928.
- (1922a) Evolutionary faith and modern doubts. Science 55: 55-61.
- (1922b) Interspecific sterility. Nature 110, 76.
- (1924) Progress in biology. In *William Bateson, F.R.S. Naturalist. His Essays and Addresses,* pp. 399-408. Edited by B. Bateson. Cambridge University Press, Cambridge, 1928.
- (1925) Huxley and evolution. Nature 115: 715-17.
Bateson, W. and Saunders, E.R. (1902) Report 1. *Reports to the Evolution Committee of the Royal Society.* Harrison, London.
Bateson, W., Saunders, E.R. and Punnett, R.C. (1904) Report 2. *Reports to the Evolution Committee of the Royal Society.* Harrison, London.
Baum, L.F. (1900) *The Wizard of Oz.* Rand McNally, Chicago.
Bell, S.J., Chow, Y.C., Ho, J.Y.K. and Forsdyke, D.R. (1998) Correlation of Chi orientation with transcription indicates a fundamental relationship between recombination and transcription. Gene 216, 285-92.
Bell, S.J. and Forsdyke, D.R. (1999a). Accounting units in DNA. J. Theor. Biol. 197, 51-61.
- (1999b). Deviations from Chargaff's second parity rule correlate with direction of transcription. J. Theor. Biol. 197, 63-76.

Belling, J. (1925) A unique result in certain species crosses. Ztschr. fur Indukt. Abstammungs. un Vererbungsl. 39, 286-8.

Belt, T. (1874) *The Naturalist in Nicaragua*. John Murray, London.

Berkhout, B. and van Hemert, F.J. (1994) The unusual nucleotide content of the HIV RNA genome results in a biased amino acid composition of HIV proteins. Nucleic Acids Res. 22, 1705-11.

Berlocher, S.H. (1998) A brief history of research on speciation. In: *Endless Forms: Species and Speciation*. pp. 3-15. Edited by D.J. Howard and S.H. Berlocher. Oxford University Press, Oxford.

Bernardi, G. and Bernardi, G. (1986) Compositional constraints and genome evolution. J. Mol. Evol. 24, 1-11.

Bernardi, G. (1993) The vertebrate genome: isochores and evolution. Mol. Biol. Evol. 10, 186-204.

Bernardi, G., Mouchiroud, D. and Gautier, C. (1993) Silent substitutions in mammalian genomes and their evolutionary implications. J. Mol. Evol. 37, 583-9.

Bernstein, C. and Bernstein, H. (1991) *Aging, Sex and DNA Repair*. Academic Press, SanDiego.

- (1997) Sexual communication. J. Theor. Biol. 188, 69-78.

Bovari, T. (1904) Cited in Stern (1950).

Bowler, P.J. (1983) *The Eclipse of Darwinism*. Johns Hopkins University Press, Baltimore.

Braun, R.E. (1998) Every sperm is sacred – or is it? Nature Genetics 18, 202-4.

Brendel, V., Dohlman, J., Blaisdell, B.E. and Karlin, S. (1991) Very long charge runs in systemic lupus erythematosus-associated autoantigens. Proc. Natl. Acad. Sci. USA 88, 1536-40.

Breslauer, K.A., Frank, R., Blocker, H. and Marky, L.A. (1986). Predicting DNA duplex stability from the base sequence. Proc. Natl. Acad. Sci. USA 83, 3746-50.

Bronson, E.C. and Anderson, J.N. (1994) Nucleotide composition as a driving force in the evolution of retroviruses. J. Mol. Evol. 38, 506-32.

Bull, J.J., Jacobson, A., Badgett, M.R. and Molineux, I.J. (1998) Viral escape from antisense RNA. Molec. Microbiol. 28, 835-46.

Bush, G.L., Case, S.M., Wilson, A.C. and Patton, J.L. (1977) Rapid speciation and chromosomal evolution in mammals. Proc. Natl. Acad. Sci. USA 74, 3942-6.

Cantor, C.R. and Schimmel, P.R. (1980). *Biophysical Chemistry*, pp. 1183-1264. Freeman, W.H., San Francisco, CA.

Carlson, E.A. (1981) *Genes, Radiation and Society. The Life and Work of H J.Muller*. Cornell University Press, Ithaca.

Catchpool, E. (1884) An unnoticed factor in evolution. Nature 31, 4.

Chandley, A.C., Jones, R.C., Dott, H.M., Allen, W.R. and Short, R.V. (1974) Meiosis in interspecific equine hybrids. 1. The male mule (*Equus asinus* X *E. caballus*) and hinny (*E. caballus* X *E. asinus*). Cytogenet. Cell Genet. 13, 330-41.

Chardin, P.T. de (1959) *The Phenomenon of Man*. Collins, London.

Chargaff, E. (1951) Structure and function of nucleic acids as cell constituents. Fed. Proc. 10, 654-9.

– (1963) *Essays on Nucleic Acids*. Elsevier, Amsterdam.

– (1979) How genetics got a chemical education. Ann. N.Y. Acad. Sci. 325, 345-60.

Chen, J-H., Le, S-Y., Shapiro, B., Currey, K.M. and Maizel, J.V. (1990) A computational procedure for assessing the significance of RNA secondary structure. CABIOS 6, 7-18.

Cohen, J. (1997) Looking for leads in HIV's battle with the immune system. Science 276, 1196-7.

Coleman, W. (1970) Bateson and chromosomes: conservative thought in science. Centaurus 15, 228-314.

Conan Doyle, A. (1890) The Sign of Four. Lippincott's Monthly Magazine (Philadelphia) 266, 172.

Cook, P.R. (1997) The transcriptional basis of chromosome pairing. J. Cell Sci. 110, 1033-40.

Coyne, J.A. (1992) Genetics and speciation. Nature 355, 511-15.

Coyne, J.A. and Orr, H.A. (1998) The evolutionary genetics of speciation. Phil. Trans. Roy. Soc. B. (Lond.) 353, 287-305.

Cox, E.C. and Yanofsky, C. (1967) Altered base ratios in the DNA of an *Echerichia coli* mutator strain. Proc. Natl. Acad. Sci. USA 58, 1895-1902.

Crick, F. (1971) General model for chromosomes of higher organisms. Nature 234, 25-27.

– (1988) *What Mad Pursuit. A Personal View of Scientific Discovery*. Harper-Collins, New York.

Cristillo, A.D., Lillicrap, T.P. and Forsdyke, D.R. (1998) Purine-loading of EBNA-1 mRNA avoids sense-antisense "collisions." FASEB J. 12, A1453 (abstract number 828).

Cristillo, A.D., Mortimer, J.R., Barrette, I.P., Lillicrap, T.P. and Forsdyke, D.R. (2001) Double-stranded RNA as a not-self alarm signal. To evade, most viruses purine-load their RNAs, but some (HTLV-1, Epstein-Barr) pyrimidine-load. J. Theor. Biol. 208, 475-91.

Cross, S.H. and Bird, A.P. (1995). CpG islands and genes. Curr. Opin. Genet. Devel. 5, 309-14.

Crowther, C.R. (1922) Evolutionary faith and modern doubts. Nature 109, 777.

Cunningham, J.T. (1922) Species and adaptations. Nature 109, 775-7.

Dang, K.D., Dutt, P.B. and Forsdyke, D.R. (1998) Chargaff differences correlate with transcription direction in the bithorax complex of *Drosophila*. Biochem. Cell Biol. 76, 129-37.

Darlington, C.D. (1932) *Recent Advances in Cytology*. J. & A. Churchill, London.

Darwin, C. (1859) *The Origin of Species by Means of Natural Selection or the Preservation of Favoured Races in the Struggle for Life*. John Murray, London.

– (1862) Notes on the causes of cross and hybrid sterility. In *The Correspondence of Charles Darwin*, Volume 10. pp. 700-711. Edited by F. Burkhardt, D.M.

Porter, J. Harvey and J.R. Topham. Cambridge University Press, Cambridge, 1997.

— (1871) *The Descent of Man and Selection in Relation to Sex*. John Murray, London.

— (1875) *The Variation of Animals and Plants under Domestication. Volume II*, 2nd Edition. John Murray, London. [The 1st edition was in 1868.]

— (1878) *The Effects of Cross and Self Fertilization in the Vegetable Kingdom*. 2nd Edition. John Murray, London. [The 1st edition was in 1877.]

Darwin, F. (1886) Physiological selection and the origin of species. Nature 34, 407.

Darwin, F. and Seward, A.C. (1903) *More Letters of Charles Darwin*. John Murray, London.

Dawkins, R. (1976) *The Selfish Gene*. Oxford University Press, Oxford.

— (1983) Universal Darwinism. In *Evolution from Molecules to Man*, Ed. by D.S. Bendall. pp. 403-25. Cambridge University Press, Cambridge.

— (1986) *The Blind Watchmaker*. Norton, London.

Delboeuf, J. (1877) Les mathématiques et le transformisme. Une loi mathématique applicable a la théorie du transformisme. La Revue Scientifique 29, 669-79.

Delbrück, M. (1949) A physicist looks at biology. Trans. Conn. Acad. Arts Sci. 38, 173-90.

— (1971) Aristotle-totle-totle. In *Of Microbes and Life*, pp. 50-55. Edited by J. Monod and F. Borek. Columbia University Press, New York.

De Vries, H. (1889) *Intracellulare Pangenesis*. First edition published by Fischer, Jena, was translated from German by C.S. Gager 1910. Open Court Pub. Co., Chicago.

Dixon, C. (1885) *Évolution without Natural Selection: or The Segregation of Species without the Aid of the Darwinian Hypothesis*. R.H. Porter, London.

Dobzhansky, T. (1937) *Genetics and the Origin of Species*. Columbia University Press, New York.

Dohlman, J.G., Lupas, A. and Carson, M. (1993) Long charge-rich alpha-helices in systemic autoantigens. Biochem. Biophys. Res. Comm. 195, 686-96.

Doyle, G.G. (1978) A general theory of chromosome pairing based on the palindromic DNA model of Sobell with modifications and amplification. J.Theor. Biol. 70, 171-84.

Du Pasquier, L., Miggiano, V.C., Kobel, H.R. and Fischberg, M. (1977) The genetic control of histocompatibility reactions in natural and laboratory-made polyploid individuals of the clawed toad *Xenopus*. Immunogenetics 5, 129-41.

Eguchi, Y., Itoh, T. and Tomizawa, J. (1991) Antisense RNA. Annu. Rev. Biochem. 60, 631-52.

Federley, H. (1913) Das Verhalten der Chromosomen bei der Spermatogenese der Schmetterlinge *Pygaera anachoreta, curtula* und *pigra* sowie einiger ihrer Bastarde. Ein Beitrag zur Frage des konstanten immediaren Art bastarden under der Spermatogenese de Pedioptera. Ztschr. fur Indukt. Abstammungs. un Vererbungsl. 9, 1-110.

Filipski, J. (1990) Evolution of DNA sequence. Contributions of mutational bias and selection to the origin of chromosomal compartments. Adv. Mutagenesis Res. 2, 1-54.

Fire, A. (1999) RNA-triggered gene silencing. Trends in Genetics 15, 358-63.

Fisher, R.A. (1931) The evolution of dominance. Biol. Rev. 6, 345-68.

– (1930) *The Genetical Theory of Natural Selection*. Oxford University Press.

Fitch, W.M. (1974) The large extent of putative secondary nucleic acid structure in random nucleic acid sequences and derived messenger-RNAs. J. Mol. Evol. 3, 279-91.

Forsdyke, D.R. (1981) Are introns in-series error-detecting codes? J. Theor. Biol. 93, 861-6.

– (1991) Early evolution of MHC polymorphism. J. Theor. Biol. 150, 451-6.

– (1993) On giraffes and peer review. FASEB. Journal 7, 619-21.

– (1994a) The heat-shock response and the molecular basis of genetic dominance. J. Theor. Biol. 167, 1-5.

– (1994b) Relationship of X chromosome dosage compensation to intracellular self/not-self discrimination: a resolution of Muller's paradox? J. Theor. Biol. 167, 7-12.

– (1994c) Do you think God is in a hurry? Globe and Mail, Toronto. October 29th.

– (1995a) Entropy-driven protein self-aggregation as the basis for self/not-self discrimination in the crowded cytosol. J. Biol. Sys. 3, 273-87.

– (1995b) Fine tuning of intracellular protein concentrations, a collective protein function involved in aneuploid lethality, sex-determination and speciation? J. Theor. Biol. 172, 335-45.

– (1995c). A stem-loop "kissing" model for the initiation of recombination and the origin of introns. Mol. Biol. Evol. 12, 949-58.

– (1995d). Conservation of stem-loop potential in introns of snake venom phospholipase A_2 genes. An application of FORS-D analysis. Mol. Biol. Evol. 12, 1157-165.

– (1995e). Relative roles of primary sequence and (G+C)% in determining the hierarchy of frequencies of complementary trinucleotide pairs in DNAs of different species. J. Mol. Evol. 41, 573-81.

– (1995f). Reciprocal relationship between stem-loop potential and substitution density in retroviral quasispecies under positive Darwinian selection. J. Mol. Evol. 41, 1022-137.

– (1995g) The origins of the clonal selection theory of immunity as a case study for evaluation in science. FASEB. J. 9, 164-6.

– (1996a). Different biological species "broadcast" their DNAs at different (G+C)% "wavelengths." J. Theor. Biol. 178, 405-17.

– (1996b). Stem-loop potential: a new way of evaluating positive Darwinian selection? Immunogenetics 43, 182-9.

- (1998a) An alternative way of thinking about stem-loops in DNA. J. Theor. Biol. 192, 489-504.
- (1999a) "The Origin of Species," revisited. A Victorian who anticipated modern developments in Darwin's theory. Queen's Quarterly 106, 112-34.
- (1999b) Haldane's rule: reproductive and species differentiations share a common path. FASEB. J. 13, A1543. Abstract #1213.
- (1999c) Two levels of information in DNA. Relationship of Romanes' "intinsic" peculiarity of the reproductive system, and Bateson's "residue," to the species-dependent component of the base composition, (C+G)%. J. Theor. Biol. 201, 47-61.
- (1999d) Heat-shock proteins as mediators of aggregation-induced "danger" signals: implications of the slow evolutionary fine-tuning of sequences for the antigenicity of cancer cells. Cell Stress and Chaperones 4, 205-10.
- (2000a) *Tomorrow's Cures Today? How to Reform the Health Research System.* Harwood Academic, Newark, NJ.
- (2000b) Haldane's rule: hybrid sterility affects the heterogametic sex first because sexual differentiation is on the path to species differentiation. J. Theor. Biol. 204, 443-52.
- (2000c) Double-stranded RNA and/or heat shock as initiators of chaperone mode switches in diseases associated with protein aggregation. Cell Stress and Chaperones 5, 375-6.
- (2001a) Functional constraint and molecular evolution. *Encyclopedia of Life Sciences.* Macmillan Reference Ltd., London.
- (2001b) Gregor Johann Mendel. *Encyclopedia of Life Sciences.* Macmillan Reference Ltd., London.
- (2001c) George John Romanes. *Encyclopedia of Life Sciences.* Macmillan Reference Ltd., London.
- (2001d) William Bateson. *Encyclopedia of Life Sciences.* Macmillan Reference Ltd., London.
- (2001e) John Burdon Sanderson Haldane. *Encyclopedia of Life Sciences.* Macmillan Reference Ltd., London.
- (2001f) Erwin Chargaff. *Encyclopedia of Life Sciences.* Macmillan Reference Ltd., London.
- (2001g) Herman J. Muller. *Encyclopedia of Life Sciences.* Macmillan Reference Ltd., London.
- (2001h) Adaptive value of polymorphism in intracellular self/not-self discrimination. J. Theor. Biol. (in press).
Forsdyke, D.R. and Mortimer, J.R. (2000) Chargaff's legacy. Gene 261, 127-37.
French, R.D. (1970) Darwin and the physiologists, or the medusa and modern cardiology. J. Hist. Biol. 3, 253-74.
Friedman, S.M., Malik, M. and Drlica, K. (1995) DNA supercoiling in a thermotolerant mutant of *Escherichia coli*. Mol Gen Genet 248, 417-22

Fulton, A.B. (1982) How crowded is the cytoplasm? Cell 30, 345-7.

Galtier, N. and Lobry, J.R. (1997) Relationships between genomic G+C content, RNA secondary structures, and optimal growth temperature in prokaryotes. J. Mol. Evol. 44, 632-6.

Gavrilov, L.A., Gavrilova, N.S., Kroutko, V.N., Evdokushkina, G.N., Semyonova, V.G., Gavrilova, A.L., Lapshin E.V., Evdokushkina, N.S. and Kushnareva, Y.E. (1997) Mutation load and human longevity. Mutation Research 377, 61-2.

Gierer, A. (1966) Model for DNA and protein interactions and the function of the operator. Nature 212, 1480-1.

Gould, S.J. (1973) Ever Since Darwin. Reflections in Natural History. W. W. Norton & Co., New York.

– (1980) Is a new and general theory of evolution emerging? Paleobiology 6, 119-30.

– (1983) The hardening of the modern synthesis. In Dimensions of Darwinism, pp. 71-93. Edited by M. Grene. Cambridge University Press, New York.

– (1998) Gulliver's further travels: the necessity and difficulty of a hierarchical theory of selection. Phil. Trans. Roy. Soc. B. (Lond.) 353, 307-14.

Grantham, R., Greenland, T., Louail, S., Mouchiroud, D., Prato, J.L., Gouy, M. and Goutier, C. (1985) Molecular evolution of viruses as seen by nucleic acid sequence study. Bull. Inst. Pasteur 83, 95-148.

Grantham, R., Perrin, P. and Mouchiroud, D. (1986) Patterns in codon usage of different kinds of species. Oxford Surveys Evol.Biol. 3, 48-81.

Green, H. and Wang, N. (1994) Codon reiteration and the evolution of proteins. Proc. Natl. Acad. Sci. USA 91, 4298-4302.

Gregory, R.P. (1905) The abortive development of the pollen in certain sweet peas. Proc. Camb. Phil. Soc. 13, 148-52.

Gulick, A. (1932) John Thomas Gulick: Evolutionist and Missionary. University of Chicago Press, Chicago.

– (1938) What are genes? 1. The Genetic and Evolutionary Picture. Quart. Rev. Biol. 13, 1-18.

Gulick, J.T. (1872a) On the variation of species as related to their geographical distribution, illustrated by the Achatinellinae. Nature 6, 222-4.

– (1872b) On diversity of evolution under one set of external conditions. J. Linn. Soc. (Zool.) 11, 496-505.

– (1887) Divergent evolution through cumulative segregation. J. Linn. Soc. (Zool.) 20, 189-274.

– (1905) Evolution, Racial and Habitudinal. Carnegie Institute, Washington, DC. Publication number 25.

Guyer, M.F. (1900) Spermatogenesis in hybrid pidgeons. Science 21, 248-9.

– (1902) Hybridism and the Germ Cell. Bull. Univ. Cincinnati 21, 1-20.

Haldane, J.B.S. (1922) Sex ratio and unidirectional sterility in hybrid animals. J. Genetics 12, 101-9.

– (1927) *Possible Worlds and Other Essays*. Chatto & Windus, London.

– (1929) The species problem in the light of genetics. Nature 124, 514-16.

– (1930) A note on Fisher's Theory of the origin of dominance and on a correlation between dominance and linkage. Am. Nat. 64, 87-90.

– (1931) Mathematical theory of natural and artificial selection. Part VII. Metastable populations. Proc. Camb. Phil. Soc. 27, 137-42.

– (1932) *The Causes of Evolution*. Longmans & Green, London.

– (1957) The theory of evolution, before and after Bateson. J. Genet. 56, 11-27.

– (1964) A defence of bean-bag genetics. Persp. Biol. Med. 7, 343-50.

Haldane, L.K. (1961) *Friends and Kindred*. Faber & Faber, London.

Hamilton, A.J. and Baulcombe, D C. (1999) A species of small antisense RNA in post-transcriptional gene silencing in plants. Science 286, 950-1.

Hamilton, W.D. (1964) The genetic evolution of social behaviour. J. Theor. Biol. 17, 1-54.

Hamming, R.W. (1980) *Coding and Information Theory*. Prentice-Hall, Englewood Cliffs.

Hare, J.T. and Taylor, J.H. (1985) One role of DNA methylation in vertebrate cells is strand discrimination in mismatch repair. Proc. Natl. Acad. Sci. 82, 7350-4.

Hawley, R.S. and Arbel, T. (1993) Yeast genetics and the fall of the classical view of meiosis. Cell 72, 301-3.

Hedrick, S M. (1992) Dawn of the hunt for nonclassical MHC function. Cell 70, 177-80.

Henikoff, S. (1997) Nuclear organization and gene expression: homologous pairing and long-range interactions. Curr. Opin. Cell Biol. 9, 388-95.

Hitchcock, D.I. (1924) Proteins and the Donnan equilibrium. Physiol. Rev. 4, 505-31.

Holliday, R. (1990) The history of heteroduplex DNA. BioEssays 12, 133-41.

Hood, L., Rowen, L. and Koop, B.F. (1995) Human and mouse T-cell receptor loci. Genomics, evolution, diversity and serendipity. Annals of the New York Academy of Sciences 758, 390-412.

Hooker, J. (1860) On the origination and distribution of species. Introductory essay on the flora of Tasmania. American Journal of Science and Arts 29, 1-25, 305-26.

– (1862a) Letter to Bates. In *The Correspondence of Charles Darwin*, Volume 10. pp. 127-30. Edited by F. Burkhardt, D.M. Porter, J. Harvey and J.R. Topham. Cambridge University Press, Cambridge, 1997.

– (1862b) Letter to Darwin. In *The Correspondence of Charles Darwin*, Volume 10. pp. 569-71. Edited by F. Burkhardt, D.M. Porter, J. Harvey and J.R. Topham. Cambridge University Press, Cambridge, 1997.

Humbert, O., Prudhomme, M., Hakenbeck, R., Dowson, C.G. and Claverys, J.P. (1995) Homeologous recombination and mismatch repair during transformation in Streptococcus pneumoniae: saturation of the Hex mismatch repair system. Proc. Natl. Acad. Sci. USA 92, 9052-6.

Hunter, L. (1993) Molecular biology for computer scientists. In *Artificial Intelligence and Molecular Biology*, pp. 1-46. Edited by L. Hunter. MIT Press, Boston.

Hunter, N., Chambers, S.R., Louis, E.J. and Borts, R.H. (1996) The mismatch repair system contributes to meiotic sterility in an interspecific yeast hybrid. EMBO. J. 15, 1726-33.

Huxley, J.S. (1942) *Evolution. The Modern Synthesis*. Allen & Unwin, London.

Huxley, T.H. (1859) The Darwinian hypothesis. In *Darwiniana. Collected Essays*. pp. 1-21. Macmillan and Co., London, 1893.

–   (1860) The origin of species. In *Darwiniana. Collected Essays*. pp. 22-79. Macmillan and Co., London, 1893.

–   (1963) A critical examination of the position of Mr. Darwin's work "On the Origin of Species," in relation to the complete theory of the causes of the phenomena of organic nature. In *Darwiniana. Collected Essays*. pp. 447-75. Macmillan and Co., London, 1893.

–   (1864) Criticisms on "The Origin of Species." In *Darwiniana. Collected Essays*. pp. 80-106. Macmillan and Co., London, 1893.

–   (1869a) *An Introduction to the Classification of Animals*. Churchill, London.

–   (1869b) The genealogy of animals. In *Darwiniana. Collected Essays*. pp. 107-19. Macmillan and Co., London, 1893.

–   (1871) Mr. Darwin's critics. In *Darwiniana. Collected Essays*. pp. 120-186. Macmillan and Co., London, 1893.

–   (1880) The coming of age of "The Origin of Species." In *Darwiniana. Collected Essays*. pp. 227-43. Macmillan and Co., London, 1893.

–   (1882) Charles Darwin. Nature 25, 597.

–   (1888) Obituary of Darwin. In *Darwiniana. Collected Essays*. pp. 253-302. Macmillan, London, 1893.

–   (1893) Preface. In *Darwiniana. Collected Essays*. Macmillan and Co., London, 1893.

–   (1896) *Evolution and Ethics and Other Essays*. Appleton, New York.

–   (1900) Letters and addresses. In *Life and Letters of Thomas Henry Huxley*. Edited by L.Huxley. Macmillan & Co., London.

Huxley, T.H. and Allen, G. (1888) *A Half Century of Science*. Fitzgerald, New York.

Iltis, H. (1932) *Life of Mendel*. George Allen & Unwin, London.

Izant, J. G. and Weintraub, H. (1984). Inhibition of thymidine kinase gene expression by anti-sense RNA: a molecular approach to genetic analysis. Cell 36, 1007-15.

Jarrett, J.T. and Lansbury, P.J. (1993) Seeding "one-dimensional crystallization" of amyloid: a pathogenic mechanism in Alzheimer's disease and amyloid? Cell 73, 1055-58.

Jenkin, F. (1867) The origin of species. The North British Review 46, 277-318.

Johannsen, W. (1911) The genotype conception of heredity. American Naturalist 45, 129-59.

Jordan, A. (1873) Remarques sur le fait d'existence en societe a l'etat sauvage des especes vegetale affine. Congres de l'Association Francaise pour l'Avancement des Sciences.

Joyce, G.F. and Orgel, L.E. (1993) Prospects for understanding the origin of the RNA world. In *The RNA World*, pp. 1-25. Edited by R. F. Gesteland and J. F. Atkins. Cold Spring Harbor Laboratory Press, New York.

Kacser, H. and Burns, J.A. (1980) The molecular basis of dominance. Genetics 97, 639-66.

Kagawa, Y., Nojima, H., Nukiwa, N., Ishizuka, M., Nakajima, T., Yasuhara, Y., et al. (1984) High G + C content in the third letter of codons of an extreme thermophile. J. Biol. Chem. 259, 2956-60.

Karkas, J.D., Rudner, R. and Chargaff, E. (1968) Separation of B. subtilis DNA into complementary strands, II. Template functions and composition as determined by transcription with RNA polymerase. Proc. Natl. Acad. Sci. USA 60, 915-20.

Karlin, S. (1995) Statistical significance of sequence patterns in proteins. Curr. Opin. Struct. Biol. 5, 360-71.

Karlin, S., Blaisdell, B.E., Mocarski, E.S. and Brendel, V. (1988) A method to identify distinctive charge configurations in protein sequences, with application to human herpesvirus polypeptides. J. Mol. Biol. 205, 165-77.

Karlin, S., Blaisdell, B.E. and Schachtel, G.A. (1990) Contrasts in codon usage of latent versus productive genes of Epstein-Barr virus: data and hypothesis. J. Virol. 64, 4264-73.

Karlin, S. and Brendel, V. (1992) Chance and statistical significance in protein and DNA sequence analysis. Science 257, 39-49.

Karlin, S. and Mrasek, J. (1996) What drives codon choice in human genes? J. Mol. Biol. 262, 459-71.

Kellogg, V.L. (1907) *Darwinism Today*. Holt, New York.

Khinchin, A.I. (1957) *Mathematical Foundations of Information Theory*. Dover, New York.

Kimura, M. (1989) The neutral theory of molecular evolution and the world view of the neutralists. Genome 31, 24-31.

King, M. (1993) *Species Evolution. The Role of Chromosome Change*. Cambridge University Press, Cambridge.

Kleckner, N. and Weiner, B.M. (1993) Potential advantages of unstable interactions for pairing of chromosomes in meiotic, somatic and premeiotic cells. Cold Spring Harbour Symp. Quant. Biol. 58, 553-65.

Kleckner, N. (1997). Interactions between and along chromosomes during meiosis. Harvey Lectures 91, 21-45.

Lankester, E.R. (1889) Review of *Darwinism* by A. R. Wallace. Nature 40, 566-70.

Lao, P.J. and Forsdyke, D.R. (2000a) Thermophilic bacteria strictly obey Szybalski's transcription direction rule and politely purine-load RNAs with both adenine and guanine. Genome Research 10, 228-36.

- (2000b) Crossover hotspot instigator (Chi) sequences in *E. coli* occupy distinct recombination/transcription islands. Gene 243, 47-57.

Larhammer, D. and Risinger, C. (1994) Why so few pseudogenes in tetraploid species? Trends. Genet. 10, 418-19.

Lauffer, M.A. (1975) *Entropy-driven Processes in Biology*. Springer-Verlag, New York, NY.

Le, S-Y. and Maizel, J.V. (1989) A method for assessing the statistical significance of RNA folding. J. Theor. Biol. 138, 495-510.

Lederberg J. and Tatum, E.L. (1946) Novel genotypes in mixed cultures of biochemical mutants of bacteria. Cold Spring. Harb. Symp. Quant. Biol. 11, 113-14.

Lesch, J.E. (1975) The role of isolation in evolution: George J. Romanes and John T. Gulick. Isis 66, 483-503.

Levitskaya, J., Coram, M., Levitsky, V., Imreh, S., Steigerwald-Mullen, P.M., Klein, G., Kurilla, M.G. and Masucci, M.G. (1995) Inhibition of antigen processing by the internal repeat region of the Epstein-Barr virus nuclear antigen-1. *Nature* 375, 685-8.

Lewis, E.B. (1954). The theory and application of a new method of detecting chromosomal rearrangements in *Drosophila melanogaster*. Am. Nat. 88, 225-39.

Lindsley, D.L., Sandler, L., Baker, B.S., Carpenter, A.T., Denell, R.E., Hall, J.C., Jacobs, P.A., Miklos, G.L., Davis, B.K., Gethmann, R.C., Hardy, R.W., Hessler, A., Miller, S.M., Nozawa, H., Parry, D.M. and Gould-Somero, M. (1972) Segmental aneuploidy and the genetic gross structure of the Drosophila genome. Genetics 71, 157-84.

Littlejohn, M.J. (1981) Reproductive isolation: a critical review. In *Evolution and Speciation: Essays in Honour of M.J.D. White*. pp. 298-334. Edited by W.A. Atchley & D.S. Woodruff. Cambridge University Press, Cambridge.

Lobry, J.R. (1996a). Asymmetric substitution patterns in the two DNA strands of bacteria. Mol. Biol. Evol. 13, 660-5.

- (1996b). Origin of replication of *Mycoplasma genitalium*. Science 272, 745-6.

Lyell, C. (1863) *The Geological Evidences of the Antiquity of Man with Remarks on Theories of the Origin of Species by Variation*. G.W. Childs, Philadelphia.

Lyon, M.F. (1992) Some milestones in the history of X-chromosome inactivation. Annu. Rev. Genet. 26, 17-28.

Macdougal, W.T. (1911) Alterations in heredity induced by ovarian treatments. Botanical Gazette 51, 241-50.

Majewski, J. and Cohan, F.M. (1998) The effect of mismatch repair and heteroduplex formation on sexual isolation in Bacillus. Genetics 148, 13-18.

Manning, C.J., Wakeland, E.K. and Potts, W.K. (1992) Communal nesting patterns in mice implicate MHC genes in kin recognition. Nature 360, 581-3.

Marchant, J. (1916) *Alfred Russel Wallace. Letters and Reminiscences*. Harper, New York.

Matassi, G., Melis, R., Macaya, G. and Bernardi, G. (1991) Compositional bimodality of the nuclear genome of tobacco. Nucleic Acids Res. 19, 5561-7.

Matic, I., Rayssiguier, C. and Radman, M. (1995) Interspecies gene exchange in bacteria: the role of SOS and mismatch repair systems in evolution of species. Cell 80, 507-15.

Maynard Smith, J. (1989) *Evolutionary Genetics*. Oxford University Press, Oxford.

Mayr, E. (1980) Some thoughts on the history of the evolutionary synthesis. In *The Evolutionary Synthesis: Perspectives on the Unification of Biology*, pp. 1-48. Edited by E. Mayr and W.B. Provine. Harvard University Press, Cambridge.

– (1982) *The Growth of Biological Thought. Diversity, Evolution, Inheritance*, pp. 564-5. Harvard University Press, Cambridge.

McConkey, E.H. (1982) Molecular evolution, intracellular organization, and the quinary structure of proteins. Proc. Natl. Acad. Sci. USA 79, 3236-40.

Melton, D.A. (1985) Injected anti-sense RNAs specifically block messenger RNA translation in vivo. Proc. Natl. Acad. Sci. USA 82, 144-8.

Mendel, G. (1865) Versuche uber Pflanzen Hybriden. Verh. naturf. Ver. in Brunn 4, 3-47.

Montgomery, T.H. (1902) A study of the chromosomes of the germ cells of metazoa. Trans. American. Phil. Soc. 20, 154-236.

Morgan, T.H. (1905) The origin of species through selection contrasted with their origin through the appearance of definite variations. Popular Science Monthly 67, 54-65.

– (1932) The rise of genetics. In *Proc. 6th Internat. Conf. Genetics*, 1, pp. 87-103. Edited by D. F. Jones. Banta, Menasha, WI.

Moroney, M.J. (1951) *Facts from Figures*. Penguin, Harmondsworth.

Mukherjee, S., Trivedi, P., Dorfman, D.M., Klein, G. and Townsend, A. (1998) Murine cytotoxic T lymphocytes recognize an epitope in an EBNA-1 fragment, but fail to lyse EBNA-1-expressing mouse cells. J. Exp. Med. 187, 445-50.

Muller, H.J. (1914) A new mode of segregation in Gregory's tetraploid primulas. American Naturalist 48, 508-512.

– (1922) Variation due to change in the individual gene. American Naturalist 56, 32-50.

– (1925) Why polyploidy is rarer in animals than in plants. Am. Nat. 59, 346-53.

– (1928) The production of mutations by X-rays. Proc. Natl. Acad. Sci. USA 14, 714-26.

– (1941) Résumé and perspective of the symposium on genes and chromosomes. Cold Spring Harbor Symposium on Quantitative Biology 9, 290-308.

– (1948) Evidence on the precision of genetic adaptation. Harvey Lect. 43, 165-229.

Murchie, A.I.H., Bowater, R., Aboul-Ela, F. and Lilley, D.M.J. (1992) Helix opening transitions in supercoiled DNA. *Biochem. Biophys. Acta* 1131, 1-15.

Murray, A.W. (1992) Creative blocks: cell cycle checkpoints and feedback controls. Nature 359, 599-604.

Muto, A. and Osawa, S. (1987) The guanine and cytosine content of genomic DNA and bacterial evolution. Proc. Natl. Acad. Sci. USA 84, 166-69.

Naumov, G.I. (1987) Genetic basis for classification and identification of the ascomycetous yeasts. Studies in Mycology 30, 469-75.

Navashin, M. (1934) Chromosomal alterations caused by hybridization and their bearing upon certain general genetic problems. Cytologia 5, 169-203.

Naveira, H.F. and Maside, X.R. (1998) The genetics of hybrid male sterility in *Drosophila*. In: *Endless Forms. Species and Speciation*. pp. 330-38. Edited by Howard, D.J. and Berlocher, S.H. Oxford University Press, Oxford.

Nicolas, A. (1998) Relationship between transcription and initiation of meiotic recombination: towards chromatin accessibility. Proc. Natl. Acad. Sci. USA 95, 87-9.

Nichols, B P., Blumenberg, M. and Yanofsky, C. (1981) Comparison of the nucleotide sequence of *trpA* and sequences immediately beyond the *trp* operon of *Klebsiella aerogenes, Salmonella typhi*, and *Escherichia coli*. Nucleic Acids Res. 9, 1743-55.

Nussinov, R. (1981) Eukaryotic dinucleotide preference rules and their implications for degenerate codon usage. J. Mol. Biol. 149, 125-31.

Ohno, S. (1967) *Sex Chromosomes and Sex-Linked Genes*. Springer-Verlag, New York.

— (1970) *Evolution by Gene Duplication*. Springer-Verlag, New York.

— (1991) To be or not to be a responder in T-cell responses: ubiquitous oligopeptides in all proteins. Immunogenetics 34, 215-21.

Olby, R. (1985) *The Origins of Mendelism*. 2nd edition. University of Chicago Press, Chicago.

— (1974) *The Path to the Double Helix*. University of Washington Press, Seattle.

Okamuro, J.K. and Goldberg, R.B. (1985) Tobacco single-copy DNA is highly homogeneous to sequences present in the genomes of its diploid progenitors. Mol. Gen. Genet. 198, 290-8.

Ordway, J.M., Tallaksen-Greene, S., Gutekunst, C-A., Bernstein, E.M., Cearley, J.A., Wiener, H.W., Dure, L.S., Lindsey, R., Hersch, S.M., Jope, R.S., Albin, R.L. and Detloff, P.J. (1997) Ectopically expressed CAG repeats cause intranuclear inclusions and a progressive late onset neurological phenotype in the mouse. Cell 91, 753-63.

Orr, H.A. (1990) "Why polyploidy is rarer in animals than in plants" revisited. Am. Nat. 136, 759-70.

— (1996) Dobzhansky, Bateson, and the genetics of speciation. Genetics 144, 1331-5.

— (1999) The devil and Darwin. Trends in Ecology and Evolution 14, 289-90.

Oshima, T., Hamasaki, N., Uzawa, T., and Friedman, S.M. (1990) Biochemical functions of unusual polyamines found in the cells of extreme thermophiles. In *The Biology and Chemistry of Polyamines*. pp 1-10. Edited by S.H. Goldembeg and I. D. Algranati. Oxford University Press, New York,

Page, A.W. and Orr-Weaver, T.L. (1996) Stopping and starting the meiotic cell cycle. Current Opinion in Genetics and Development 7, 23-31.

Prabhu, V.V. (1993) Symmetry observations in long nucleotide sequences. Nucleic Acids Res. 21, 2797-800.

Presgraves, D.C. and Orr, H.A. (1998) Haldane's rule in taxa lacking a hemizygous X. Science 282, 952-4.

Provine, W.B. (1971) *The Origins of Theoretical Population Genetics*. The University of Chicago Press, Chicago.

– (1986) *Sewell Wright and Evolutionary Biology*. University of Chicago Press, Chicago.

– (1992) Progress in evolution and the meaning of life. In *Julian Huxley, Biologist and Statesman of Science*, pp. 165-80. Edited by C.K. Waters and A. van Helden. Rice University Press, Houston.

Prusiner, S.B. (1997) Prion diseases and the BSE crisis. Science 278, 245-51.

Punnett, R.C. (1950) Early days of genetics. Heredity 4, 1-10.

Radman, M. and Wagner, R. (1993). Mismatch recognition in chromosomal interactions and speciation. Chromosoma 102, 369-73.

Radman, M., Wagner, R. and Kricker, M.C. (1993) Homologous DNA interactions in the evolution of gene and chromosome structure. Genome Anal. 7, 139-54.

Ralston, H.J. (1944) G. J. Romanes on the excitability of muscle. Science 100, 123-4.

Rayssiguier, C., Thaler, D. and Radman, M. (1989) The barrier to recombination between *Escherichia coli* and *Salmonella typhimurium* is disrupted in mismatch-repair mutants. Nature 342, 396-401.

Reeder, R.H. (1985) Mechanisms of nucleolar dominence in animals and plants. J. Cell Biol. 101, 2013-16.

Reich, Z., Wachtel, E.J. and Minsky, A. (1995) *In vivo* quantitative characterization of intermolecular interactions. J. Biol. Chem. 270, 7045-6.

Ringereide, M. (1979) Romanes – Father and son. The Bulletin of the Committee on Archives and History of the United Church of Canada 28, 35-46.

Roberts, H.F. (1929) *Plant Hybridization before Mendel*. Hafner, New York.

Robertson, H.D. and Mathews, M.B. (1996) The regulation of the protein kinase PKR by RNA. Biochimie 78, 909-14.

Rocco, V. and Nicolas, A. (1996) Sensing of DNA non-homology lowers the initiation of meiotic recombination in yeast. Genes to Cells 1, 645-61.

Rocha, E.P.C., Danchin, A. and Vairi, A. (1999) Universal replication biases in bacteria. Mol Microbiol. 32, 11-16.

Romanes, E. (1894) Letter to Mr. Huxley, 24th September. Huxley Archive at Imperial College of Science, Technology and Medicine, London.

– (1895) Letter to Mrs. Huxley, 16th July. Huxley Archive at Imperial College of Science, Technology and Medicine, London.

– (1896) *The Life and Letters of George John Romanes*, Longmans, Green & Co., London.

- (1902) *The Hallowing of Sorrow*. 4th Edition. p. 68. Longmans, Green & Co., London.
- (1907) *The Story of Port Royal*. Dutton, New York.
- (1918) *The Story of an English Sister*. Longmans, Green & Co., London.

Romanes, G. J. (1874) Rudimentary organs. Nature 9, 441-2. Disuse as a reducing cause in species. Nature 10, 164.
- (1878) *A Candid Examination of Theism*. Truebner & Co., London.
- (1880) A speculation regarding the senses. Nature 21, 348.
- (1881) *Animal Intelligence*. Kegan Paul & Trench, London.
- (1882) Charles Darwin. Nature 26, 49-51, 73-5, 97-100, 145-7, 169-71.
- (1885a) *Mind and Motion and Monism*. Longmans, Green & Co., London.
- (1885b) *Jelly-Fish, Star-Fish and Sea-Urchins: Being a Research of Primitive Nervous Systems*. Appleton, New York.
- (1885c) Evolution without natural selection. Nature 33, 26-7.
- (1886a) Physiological selection: an additional suggestion on the origin of species. Nature 34, 314-16, 336-40, 362-65,
- (1886b) Physiological selection: an additional suggestion on the origin of species. J. Linn. Soc. (Zool.) 19, 337-411.
- (1886c) Letter to Mendola. Archives of Newham Borough, London.
- (1887) Physiological selection. Nineteenth Century 21, 59-80.
- (1888a) Definition of the theory of natural selection. Nature 38, 616-18.
- (1888b) Mr. Dyer on physiological selection. Nature 39, 103-4.
- (1888c) Natural selection and the origin of species. Nature 39, 173-5.
- (1890) Wallace on physiological selection. The Monist 1, 1-20.
- (1892) *Darwin, and After Darwin: 1. The Darwinian Theory*. Longmans, Green & Co., London.
- (1893) *An Examination of Weissmanism*. Open Court Pub. Co., Chicago.
- (1895a) *Darwin, and After Darwin: 2. Post Darwinian Questions. Heredity and Utility*. Longmans, Green & Co., London.
- (1895b) *Thoughts on Religion*. Open Court Pub. Co., Chicago.
- (1897) *Darwin, and After Darwin: 3. Isolation and Physiological Selection*. Longmans, Green & Co., London.

Romano, T.M. (1993) *Making Medicine Scientific. John Burdon Sanderson and the Culture of Victorian Science*. Ph.D. thesis. Yale University, New Haven.
- (1997) The cattle plague of 1865 and the reception of "the germ theory" in mid-Victorian England. J. Hist. Med. Allied. Sci. 52, 51-80.

Roser, B., Brown, R.E. and Singh, P.B. (1991) Excretion of transplantation antigens as signals of genetic individuality. In *Chemical Senses. 3. Genetics of Perception and Communications*, pp. 187-209. Edited by C. J. Wysocki and M. R. Kare. Marcel Dekker, Basel.

Russell, W. (1890) A characteristic organism of cancer. Brit. Med. J. 2, 1356-60.

Rynditch, A.V., Zoubek, S., Tsyba, L., Tryapitsina-Guley, N. and Bernardi, G. (1998) The regional integration of retroviral sequences into the mosaic genome of mammals. Gene 222, 1-16.

Salisbury, Lord (1894) Inaugural address of the Most Hon. the Marquis of Salisbury, K.G., D.C.L., F.R.S., Chancellor of the University of Oxford, President. Nature 50, 339-43.

Salser, W. (1970) Discussion. Cold Spring Harb. Quant. Biol. 35, 19.

– (1978) Globin mRNA sequences: analysis of base pairing and evolutionary implications. Cold Spring Harb. Symp. Quant. Biol. 42, 985-1002.

Sandler, I. (2000) Development: Mendel's legacy to genetics. Genetics 154, 7-11.

Schachtel, G.A., Bucher, P., Mocarski, E.S., Blaisdell, B.E. and Karlin, S. (1991) Evidence for selective evolution in codon usage in conserved amino acid segments of human alphaherpesvirus proteins. J. Mol. Evol. 33, 483-94.

Schild, H., Rötzschke, O., Kalbacher, H. and Rammensee, H. G. (1990) Limit of T-cell tolerance to self proteins by peptide presentation. Science 247, 1587-89.

Schwartz, J.S. (1985) George John Romanes' defence of Darwinism: the correspondence of Charles Darwin and his chief disciple. J. Hist. Biol. 28, 281-316.

Seffens, W. and Digby, D. (1999) mRNAs have greater negative folding free energies than shuffled or codon-choice randomized sequences. Nucleic Acids Res. 27, 1578-184.

Sen, G.C. and Lengyel, P. (1994) The interferon system. A bird's eye view of its biochemistry. J. Biol. Chem. 267, 5017-20.

Shannon, C.E. (1948) The mathematical theory of communication. Bell Syst. Tech. J. 27, 397-423.

Sharpey-Shafer, E. (1972) History of the Physiological Society during its First Fifty Years 1876-1927. Cambridge University Press, Cambridge, pp. 32-36.

Shaw, G.B. (1913) Pygmalion. In Bernard Shaw. Complete Plays with Prefaces. Volume I. Dodd, Mead & Co., New York, 1963.

Silverstein, A. M. (1989) A History of Immunology. Academic Press, San Diego, CA. pp. 142-5.

Sinden, R. (1994) DNA Structure and Function. Academic Press, San Diego.

Smith, K. N. and Nicolas, A. (1998) Recombination at work in meiosis. Current Opinion in Genetics and Development 8, 200-211.

Smithies, O., Engels,, W.R., Devereux, J.R., Slightom, J.L. and Shen, S. (1981) Base substitutions, length differences and DNA strand asymmetries in the human Gλ and Aλ fetal globin gene region. Cell 26, 345-53.

Smithies, O. and Powers, P.A. (1986) Gene conversions and their relations to homologous chromosome pairing. Phil. Trans. R. Soc. B. 312, 291-302.

Sobel, D. (2000) Galileo's Daughter. Fourth Estate, London.

Sobell, H.M. (1972) Molecular mechanism for genetic recombination. Proc. Natl. Acad. Sci. USA 69, 2483-7.

Srivastava, P.K., Menoret, A., Basu, S., Binder, R.J. and McQuade, K.L. (1998) Heat shock proteins come of age: primitive functions acquire new roles in an adaptive world. Immunity 8, 657-65.

Stadler, L.J. (1929) Chromosome number and the mutation rate in *Avena* and *Triticum*. Proc. Natl. Acad. Sci. USA 15, 876-81.

– (1932) On the genetic nature of induced mutation in plants. In *Proc. 6th Internat. Conf. Genetics*, 1, pp. 274-94. Edited by D.F. Jones. Banta, Menasha, WI.

Stebbins, G.L. (1980) Botany and the synthetic theory of evolution. In *The Evolutionary Synthesis. Perspectives on the Unification of Biology*, pp. 139-52. Edited by E. Mayr & W.B. Provine. Harvard University Press, Cambridge.

Stebbins, G.L. and Ayala, F. G. (1981) Is a new evolutionary synthesis necessary? Science 213, 967-71.

Stern, C. (1950) Boveri and the early days of genetics. Nature 166, 446.

Strick, T.R., Croquette, V. and Bensimon, D. (1998) Homologous pairing in stretched supercoiled DNA. Proc. Natl. Acad. Sci. USA 95, 10579-83.

Sturtevant, A.H. (1913) The linear arrangement of six sex-linked factors in Drosophila, as shown by their mode of association. J. Exp. Zool. 14, 43-59.

Sueoka, N. (1961a) Correlation between base composition of deoxyribonucleic acid and amino acid compositional of protein. Proc. Natl. Acad. Sci. USA 47 1141-9.

– (1961b) Compositional correlation between deoxyribonucleic acid and protein. Cold Spring Harbor Symp. Quant. Biol. 26, 35-43.

– (1962) The genetic basis of variation and heterogeneity of DNA base composition. Proc. Natl. Acad. Sci. USA 48, 582-92.

– (1995) Intrastrand parity rules of DNA base composition and usage biases of synonymous codons. J. Mol. Evol. 40, 318-25.

Summers, H., Fleming, A. and Frappier, L. (1997) Requirements for Epstein-Barr nuclear antigen 1 (EBNA-1)-induced permanganate sensitivity of the Epstein-Barr latent origin of DNA replication. J. Biol. Chem. 272, 26434-40.

Sutton, W. S. (1902) On the morphology of the chromosome group in *Brachystola magna*. Biological Bulletin 4, 24-35.

– (1903) The chromosomes in heredity. Biological Bulletin 4, 231-51.

Swift, J. (1733) On poetry, a rhapsody. In *Jonathan Swift, the Complete Poems*, pp. 522-36. Edited by P. Rogers. Yale University Press, New Haven, 1982.

Szathmary, E. (1999) The origin of the genetic code. Amino acids as cofactors in the RNA world. Trends in Genetics 15, 223-9.

Szostak, J.W., Orr-Weaver, T.L. and Rothstein, R.J. (1983) The double-strand-break repair model for recombination. Cell 33, 25-35.

Szybalski, W., Kubinski, H. and Sheldrick, P. (1966) Pyrimidine clusters on the transcribing strands of DNA and their possible role in the initiation of RNA synthesis. Cold Spring Harbor Symp. Quant. Biol. 31, 123-7.

Tan, C. C. (1935) Salivary gland chromosomes in the two races of *Drosophila pseudoobscura*. Genetics 20, 392-402.

Templeton, A.R. (1989) The meaning of species and speciation: a genetic perspective. In *Speciation and its Consequences*, pp. 3-27. Edited by D. Otte and J. Endler, Sinauer Associates, Sunderland, MA.

Thiselton-Dyer, W. T. (1888a) Opening address to the British Association, Section D, Nature 38, 473-80.

– (1888b) Mr. Romanes' paradox. Nature 39, 7-9.

– (1888c) Mr. Romanes on the origin of species. Nature 39, 126-7.

Thorley-Lawson, D.A., Miyashita, E.M. and Kahn, G. (1996) Epstein-Barr virus and the B cell: That's all it takes. *Trends Microbiol.* 4, 204-7.

Tian, B., White, R.J., Xia, T., Welle, S., Turner, D.H., Mathews, M.B. and Thornton, C.A. (2000) Expanded CUG repeat RNAs form hairpins that activate the double-strand RNA-dependent protein kinase PKR. RNA 6, 79-87.

Tomizawa, J. (1984) Control of ColE1 plasmid replication: the process of binding of RNA I to the primer transcript. Cell 38, 861-70.

Turner, F.M. (1974) George John Romanes: from faith to faith. In *Between Science and Religion*. pp. 134-163. Yale University Press, New Haven.

Turner, D.H. (1996) Thermodynamics of base pairing. Curr. Opin. Struct. Biol. 6, 299-304.

Turner, D.H., Sugimoto, N. and Freier, S.M. (1988) RNA structure prediction. Annu. Rev. Biophys. Chem. 17, 167-192.

Vidovik, D. and Matzinger, P. (1988) Unresponsiveness to a foreign antigen can be caused by self-tolerance. Nature 336, 222-5.

Vorzimmer, P. (1963) Charles Darwin and the blending inheritance. Isis 54, 371-90.

Vulic, M., Dionisio, F., Taddei, F. and Radman, M. (1997) Molecular keys to speciation: DNA polymorphism and the control of genetic exchange in enterobacteria. Proc. Natl. Acad. Sci. USA 94, 9763-7.

Wagner, R.E. and Radman, M. (1975) A mechanism for initiation of genetic recombination. Proc. Natl. Acad. Sci. USA 72, 3619-22.

Wallace, A.R. (1858) On the tendency of varieties to depart indefinitely from the original type. Reprinted in *Darwin and His Critics*, pp. 16-21. Edited by B.R. Kogan. Wadsworth, San Francisco, 1960.

– (1866) *The Scientific Aspect of the Supernatural*, Farrah, F., London.

– (1869) Review of *Principles of Geology* by C. Lyell, Quarterly Review 126, 187-205.

– (1874) *On Miracles and Modern Spiritualism*. Burns, J., London.

– (1882) *Land Nationalization its Necessity and its Aims*. Trubner & Co., London.

– (1886) Romanes *versus* Darwin. Fortnightly Review 46, 300-316.

– (1889) *Darwinism*. Macmillan & Co., London.

– (1898) *Vaccination a Delusion*. Swan, Sonnenshein & Co., London.

– (1905) *My Life*, Chapman & Hall, London. Chapter 36.

Watson, J. D. and Crick, F. H. C. (1953) Genetical implications of the structure of deoxyribonucleic acid. Nature 171, 964-7.

Weismann, A. (1875) Saison-Dimorphismus der Schmetterlinge. In *Studies in the Theory of Descent.* pp. 634-44. Translated by R. Meldola. Sampson, Low, London, 1882.

– (1893) *The Germ Plasm. A Theory of Heredity.* Walter Scott, London.

– (1904) *The Evolution Theory,* Vol. 2, translated by J.A. Thomson. Arnold, London, pp. 337-9.

Wells, R.D. (1996). Molecular basis of genetic instability of triplet repeats. J. Biol. Chem. 271, 2875-8.

White, M.J.D. (1978) *Modes of Speciation.* W.H. Freeman, San Francisco.

Winge, Ö. (1917) The Chromosomes, their number and general importance. Compt. Rend. Trav. Lab. Carlsberg. 13, 131-275.

Williams, G.C. (1966) *Adaptation and Natural Selection.* Princeton University Press, Princeton.

– (1975) *Sex and Evolution.* Princeton University Press, Princeton.

– (1992) *Natural Selection: Domains, Levels, and Challenges.* Oxford University Press, Oxford.

Wong, B.C., Chiu, S-K. and Chow, S.A. (1998) The role of negative superhelicity and length of homology in the formation of paranemic joints promoted by RecA protein. J. Biol. Chem. 273, 12120-7.

Wright, S.G. (1914) Duplicate genes. American Naturalist 48, 638-9.

– (1939) *Statistical Genetics in Relation to Evolution.* Hermann & Co., Paris.

– (1968) *Evolution and the Genetics of Populations. 1. Genetic and Biometric Foundations.* University of Chicago Press, Chicago.

– (1969) *Evolution and the Genetics of Populations. 2. The Theory of Gene Frequencies.* University of Chicago Press, Chicago.

– (1977) *Evolution and the Genetics of Populations. 3. Experimental Results and Evolutionary Deductions.* University of Chicago Press, Chicago.

– (1978) *Evolution and the Genetics of Populations. 4. Variability Within and Among Natural Populations.* University of Chicago Press, Chicago.

Wyatt, G.R. (1952) The nucleic acids of some insect viruses. J. Gen. Physiol. 36, 201-89.

Xu, L. and Kleckner, N. (1995) Sequence non-specific double-strand breaks and interhomolog interactions prior to double-strand break formation at a meiotic recombination hot-spot in yeast. EMBO. J. 14, 5115-28.

Yates, J. L. and Camiolo, S.M. (1988) Dissection of DNA replication and enhancer activation functions of Epstein-Barr virus nuclear antigen 1. *Cancer Cells 6,* 197-205.

Yin, P.D. and Hu, W-S. (1997) RNAs from genetically distinct retroviruses can copackage and exchange genetic information. J. Virol. 71, 6237-42.

Zahrt, T.C. and Maloy, S. (1997) Barriers to recombination between closely related bacteria: MutS and RecBCD inhibit recombination between

Salmonella typhimurium and *Salmonella typhi*. Proc. Natl. Acad. Sci. USA 94, 9786-91.

Zehl, C. and Bell, G. (1997) The advantage of sex in evolving yeast populations. Nature 388, 465-71.

Ziman, J. (1996) Is science losing its objectivity? Nature 382, 751-4.

Zimmerman, S.B. amd Murphy, L.D. (1996) Macromolecular crowding and the mandatory condensation of DNA in bacteria. FEBS Lett. 390, 245-8.

Zuker, M. (1994) Prediction of RNA secondary structure by energy minimization. Meth. Molec. Biol. 25, 267-94.

Index